Sociable Cities

SOCIABLE CITIES

THE LEGACY OF EBENEZER HOWARD

Peter Hall

and

Colin Ward

JOHN WILEY & SONS

Chichester • New York • Weinheim • Brisbane • Singapore • Toronto

Copyright © 1998 by Peter Hall and Colin Ward

Published 1998 by John Wiley & Sons Ltd,
Baffins Lane, Chichester, West Sussex PO19 1UD, England

National 01243 779777. International (+44) 1243 779777

e-mail (for orders and customer service enquiries): cs-books@wiley.co.uk

Visit our Home Page on http://www.wiley.co.uk or http://www.wiley.com

Peter Hall and Colin Ward have asserted their right under the Copyright, Designs and Patents Act, 1988,
to be identified as the authors of this work.

OTHER WILEY EDITORIAL OFFICES

John Wiley & Sons, Inc., 605 Third Avenue, New York, NY 10158-0012, USA

WILEY-VCH Verlag GmbH, Pappelallee 3, D-69469 Weinheim, Germany

Jacaranda Wiley Ltd, 33 Park Road, Milton, Queensland 4064, Australia

John Wiley & Sons (Asia) Pte Ltd, 2 Clementi Loop #02-01, Jin Xing Distripark, Singapore 129809

John Wiley & Sons (Canada) Ltd, 22 Worcester Road, Rexdale, Ontario M9W 1L1, Canada

LIBRARY OF CONGRESS CATALOGING-IN-PUBLICATION DATA

Hall, Peter Geoffrey.
  Sociable cities : the legacy of Ebenezer Howard / Peter Hall,
Colin Ward.
    p. cm.
  Includes bibliographical references and index.
  ISBN 0-471-98504-X. — ISBN 0-471-98505-8
  1. Howard, Ebenezer, Sir, 1850–1928. 2. Garden cities. 3. New
towns. 4. City planning. I. Ward, Colin. II. Title.
  HT161.H35 1998
  307.76'8—dc21   98–4175
                                                      CIP

BRITISH LIBRARY CATALOGUING IN PUBLICATION DATA

A catalogue record for this book is available from the British Library

ISBN 0-471-98504-X (hardback)
ISBN 0-471-98505-8 (paperback)

Typeset in 9/12pt Caslon 224 from authors' disks by Mayhew Typesetting, Rhayader, Powys
Printed and bound in Great Britain by Bookcraft (Bath) Ltd.

This book is printed on acid-free paper responsibly manufactured from sustainable forestry, in which at
least two trees are planted for each one used for paper production.

CONTENTS

Foreword   vii
Graeme Bell, Town and Country Planning Association

Preface   ix

PART I: THE FIRST CENTURY

Chapter 1   Howard's Beginning   3

Chapter 2   Garden City: Ideal and Reality   17

Chapter 3   From Garden Cities to New Towns   41

Chapter 4   Plotlands: The Unauthorised Version   71

Chapter 5   Land Settlement: The Failed Alternative   79

Chapter 6   A Continental Interlude   87

PART II: THE COMING CENTURY

Chapter 7   Then and Now   103

Chapter 8   The Quest for Sustainability   123

Chapter 9   Sustainable Social Cities of Tomorrow   151

Chapter 10   Making It Happen   171

Chapter 11   Do-it-Yourself New Towns   191

Chapter 12   Not Counting NIMBIES   199

List of Figures and Tables   211

References   213

Index   223

FOREWORD

GRAEME BELL
Director, Town and Country Planning Association

Publication of *Sociable Cities – the Legacy of Ebenezer Howard* is the first celebratory event in the centenary year of the Town and Country Planning Association, which Howard and a handful of like-minded enthusiasts founded in London on June 10th 1899 as the Garden City Association. *Sociable Cities* appears 100 years to the month after *Howard's To-morrow! A Peaceful Path to Real Reform*, which launched the Garden City idea and is arguably the seminal planning text for the 20th century.

The Association's original intention was to prepare a new version of that famous book. But the authors Sir Peter Hall (who is our Chairman) and Colin Ward (formerly our Environmental Education Officer), both distinguished writers and commentators on social and environmental planning matters, have done something more interesting.

Naturally in a commemorative volume they assess Howard and the successes and failures his ideas have had in the current century, but then in a way that he would surely have appreciated, they test the relevance of his thinking to modern circumstances. We are not surprised by their conclusion that "while the world has changed out of all recognition, his message still has a startling, almost surreal, relevance to us in the Britain of 1998". Interpreting Howard's work in the light of the next century's sustainable development imperatives, they show how multi-centred sociable cities could be planned.

We hope that their application of these ideas to particular areas of central and southern Britain will stimulate widespread debate both in our councils and in public. We are determined to contribute significantly to a new consensus about how conservation and development can be reconciled in the light of the full range of society's needs – environmental, social and economic. Our concern remains as it was in Howard's day: the pursuit of social justice, prosperity and equity combined with environmental quality through practical, realisable planning policies.

As the TCPA enters its second century, we are conscious of following in the footsteps of many eminent and concerned Association members and supporters who have struggled, as we do today, to meet the ever-evolving problems of town and country planning with imagination and creativity. Our predecessors have provided enduring foundations for us to build upon.

PREFACE

Ebenezer Howard's *To-Morrow! A Peaceful Path to Real Reform* was published in October 1898. Eight months later, in June 1899, Howard founded the Garden City Association to propagate the ideas in his book. In its original edition *To-Morrow!* sold a few hundred copies, but – republished as *Garden Cities of To-Morrow* in 1902 – it was destined to become the most influential and important book in the entire history of twentieth-century city planning. Over the following 70 years it effectively produced the pioneer garden cities at Letchworth and Welwyn, in which Howard took a personal role, and then – long after his death in 1928 – some 30 new towns in the United Kingdom as well as countless imitations all over the world.

The Garden City Association became first the Garden Cities and Town Planning Association and then the Town and Country Planning Association. In this book, published to commemorate the TCPA's centenary year, we try to do two things: in Part I, to tell the story of the first hundred years of Howard's movement; in Part II, to suggest how his ideas are still completely relevant to the creation of civilised and sustainable new communities for the coming century.

Some parts of this account have seen the light of day earlier: in the pages of the TCPA's own magazine *Town and Country Planning*; in the Denman Lecture which Peter Hall gave at the Department of Land Economy, University of Cambridge, in 1996;[1] and in the report of the TCPA–Joseph Rowntree inquiry into land for housing, *The People – Where Will They Go?*[2] We are grateful to the Joseph Rowntree Foundation, and to the Department of Land Economy, for helping to develop these ideas.

The historical account in Chapters 1–3 draws heavily upon two standard works: Robert Beevers' biography of Ebenezer Howard, and Dennis Hardy's monumental history of the Town and Country Planning Association.[3] We indicate our debt to these sources many times in our footnotes, but we wish to pay special acknowledgement to them here.

We also want to acknowledge the help of Magda Hall in drawing our attention to the success of the Farmers' Market movement in California, which forms an important plank in the structure of the final chapter.

---

[1] Hall 1996.
[2] Breheny and Hall 1996c.
[3] Beevers 1988; Hardy 1991a, 1991b.

x
—

Finally, thanks to Tristan Palmer at Wiley for his ever-efficient handling of publication matters, to Peter Lamb for the maps, and to Karin Fancett for her meticulous editing of the final typescript.

PETER HALL
COLIN WARD
London and Kersey Uplands, March 1998

PART I

THE FIRST CENTURY

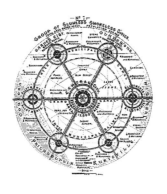

CHAPTER 1
HOWARD'S BEGINNING

When Ebenezer Howard published *To-Morrow! A Peaceful Path to Real Reform*, in October 1898, he was an obscure 48-year-old shorthand writer living in genteel poverty with a harassed wife and four children in a modest house in north London. To get the book published, he had to accept a £50 loan from George Dickman, an American who was managing director of the Kodak photographic company in Britain and a fellow-believer in spiritualism. Howard may well have taken most of the copies himself, for distribution to friends. But it sold well enough to encourage the publishers, Swan Sonnenschein, to issue a cheap paperback at a shilling, and soon after 1900 more than 3000 copies, in both editions, had been sold.[1]

It was a modest start. But, within a decade, republished as *Garden Cities of To-Morrow* in 1902, Howard's book had created an intellectual shock-wave that reverberated around the world: the first Garden City had been launched at Letchworth in Hertfordshire in 1903; German examples were already on the drawing boards, and would soon be realised; translations of the book were appearing in language after language. Half a century after its first publication, and two decades after its author's death, this modest book had spawned an Act of Parliament and the designation of a score of new towns in his native country. Seldom in history can any book have had such an extraordinary impact.

HOWARD: THE PERSON

Still, in 1898, *Tomorrow!*'s publication was an inauspicious event. Few, save perhaps Howard himself, could have believed that it would be one of those few books that, unaided, would change the course of history. And Howard categorically lacked the kind of personality that would cause him to be described a century later as a media person; he was not exactly built for photo-opportunities. Frederic J. Osborn, his faithful lieutenant and follower, said of him:

> Howard's personality was a continual source of surprise to strangers knowing of his astonishing achievements. He was the mildest and most unassuming of men, unconcerned with his personal appearance, rarely giving evidence of the force within him. Of medium height and sturdy build, and always dressed in a rather shabbily

---

[1] Fishman 1977, 54, quoting Howard's draft of an unfinished autobiography; Beevers 1988, 43, 57, 104.

conventional way, he was the sort of man who could easily pass unnoticed in a crowd; Mr. Bernard Shaw, who much admired what he did, only overstates a truth when he says that this "amazing man" seemed an "elderly nobody", "whom the Stock Exchange would have dismissed as a negligible crank".[2]

Cecil Harmsworth (brother of Alfred of the *Daily Mail*), who knew him well, described him as "our little kind friend, who moved so unassumingly amongst us".[3] A local Hertfordshire solicitor, whom he met during the purchase of the land for the first garden city at Letchworth, was less kind: "an insignificant little man who, to a piercing eye, did not appear to be worth many shillings".[4]

And yet, there was something more. He could command a platform, and just possibly he might have commanded a television studio (Figure 1):

> His most distinguished physical characteristics were a clear fresh complexion, a fine aquiline profile, and a really beautiful and powerful speaking voice, and it is not surprising that he was much in demand as an amateur Shakespearean actor in his younger days.[5]

Dr Parker of the City Temple told him he could have been a successful preacher. Though he dominated on the public platform, in private and business life he got ignored, partly because he was too preoccupied and had few concerns with administrative details; this was a product of his power of concentration. But everyone liked him, above all children.[6]

His life until the late 1890s had been one of hard grind and personal failure. Born to middle-middle-class tradespeople in the City of London on 29 January 1850 (Figure 2), he spent his childhood in small country towns in southern England: Sudbury, Ipswich, Cheshunt – a fact that may help explain his passionate love of the countryside. He left school at 15, and became a clerk in the City. But at 21, he emigrated to America to become a pioneer farmer in Nebraska. The experience was a disaster, and a year later he found himself a shorthand writer in Chicago, established in the job he was to follow all his life.

He spent four years in Chicago, 1872–1876, and they must have been formative ones. Howard always denied that he found inspiration in the windy city, but all subsequent chroniclers of his life agree that he must have got the germ for his idea of the Garden City here, in his lodgings on Michigan Avenue; Chicago itself had been known as "the garden city" before the great fire of 1871, though it would soon lose that character in the rebuilding. He must have been acquainted with the new garden suburb of Riverside on the Des Plaines river, 9 miles (14 km) outside the city, designed by the great landscape architect Frederick Law Olmsted.[7]

Then, in 1876, he returned to London. He obtained a position with Gurneys, the official Parliamentary reporters, and after one unsuccessful attempt at a

---

[2] Osborn 1946, 22–23.
[3] Thomas 1983, 1.
[4] Quoted in Beevers 1988, 86.
[5] Osborn 1946, 23.
[6] Osborn 1946, 23.
[7] Beevers 1988, 7; Osborn 1950, 226–227; Stern 1986, 133–134.

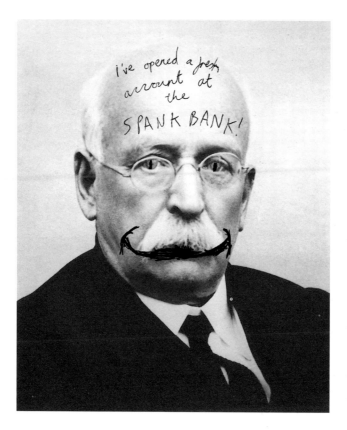

Figure 1 *Ebenezer Howard. This picture, displaying his passion and determination, must have been taken at about the time of his audacious unilateral bid for the land at Welwyn, when his friends had to bail him out (see Chapter 3).* Source: Town and Country Planning Association.

private partnership he remained with Gurneys and others in the same business for the rest of his days; "His life was always one of hard work and little income".[8] But there was one advantage, which he must have seen very clearly: he said himself that he was prepared for his work through his profession, which exposed him to argument on the major issues of the day.[9] He also had a hobby, which like many hobbies amounted to an obsession: he thought of himself as an inventor, and when obsessed with an idea he would persist with it in defiance of all advice from his friends.[10] He thought of garden cities as an invention, just as he did of an improved variable-spacing typewriter, which never came to anything.[11] And this is an important clue to his character: Osborn, who knew him so well, said, "Howard –

[8] Osborn 1946, 19.
[9] Beevers 1988, 7.
[10] Osborn 1946, 19.
[11] Beevers 1988, 12.

Figure 2 *Ebenezer Howard Plaque. The birthplace in London's Barbican; but his formative influences were Chicago and the American prairies.* Source: Peter Hall photograph.

let me emphasize this – was not a political theorist, not a dreamer, but an inventor", a man who thought of an idea and then worked out its design on paper.[12]

CITY IN FERMENT: HOWARD'S INTELLECTUAL EDUCATION

But for the 30-year-old Howard, London was more than a place to slave at shorthand and dream of a new typewriter. For, as the 1870s gave way to the 1880s, this was a city in social and intellectual turmoil.

The entire city was a hotbed of radical activities and "causes".[13] William Morris broke with H.M. Hyndman and founded *Commonweal* for the Socialist League; the anarchists produced *Freedom*, under the patronage of Prince Peter Kropotkin; and another journal, *To-Day*, was run by Henry Champion and Hubert Bland.[14] Everyone in these groups wanted a new social order, but no one quite knew what:

---

12 Osborn 1946, 21.
13 Hardy 1991a, 30.
14 MacKenzie and MacKenzie 1977, 76–77.

Should one take to the streets like Hyndman, to the stump like Morris, to the anarchist commune, the producer cooperative and the self-improving idealism of the Fellowship of the New Life?[15]

The Fellowship of the New Life, founded by the itinerant visionary Thomas Davidson together with Edward Pease and Percival Chubb in 1883, was essentially a group of people who were trying to achieve an earthly paradise in substitution for a heavenly one. Essentially, its first meeting on 24 October 1883 was the beginning of the Fabian Society.[16]

A decade later, in 1893, J. Bruce Wallace, subsequently one of the founder-members of the Garden City Association, founded his own Brotherhood Church, and its branches became a magnet for every kind of alternative cause:

Every kind of "crank" came and aired his views on the open platform, which was provided every Sunday afternoon. Atheists, Spiritualists, Individualists, Communists, Anarchists, ordinary politicians, Vegetarians, Anti-Vivisectionists and Anti-Vaccina-tionists – in fact, every kind of "anti" had a welcome and a hearing and had to stand a lively criticism in the discussion which followed.[17]

Towards the end of 1879 Howard joined a debating society, the Zetetical Society, which mainly attracted freethinkers; it already included George Bernard Shaw and Sidney Webb, with whom he was soon on good terms. During these years Howard read widely, as Osborn testifies; in his book he cited over 30 writers ranging from William Blake to the medical officer of health for Derbyshire; in particular he read newspaper reports, Royal Commission evidence, weighty articles in the *Fortnightly Review*, the Fabian Essays and the writings of J.S. Mill and Herbert Spencer, the last-mentioned first read in Chicago and perhaps the most important influence on him. He was also a regular attender at Dissenting chapels, though by now losing his faith, and much of this reading comes from a common dissenting tradition.[18] "All the chief contributors to the stock of ideas from which Howard distilled the concept of the garden city – not only Thomas Spence, the land reformer, but also Richardson, Spencer and even Henry George – were dissenters by upbringing or steeped in its tradition."[19] Apart from Kropotkin, no continental figure seems to have reached him, not even Marx.[20]

THE GREAT 1880s DEBATE: THE LAND QUESTION

By the late 1880s Howard had begun to focus on the land question. And that was no surprise, because land was one of the two or three most intensely debated topics of the age. The basic reason was that British agriculture was in deep

[15] MacKenzie and MacKenzie 1977, 77.
[16] MacKenzie and MacKenzie 1977, 15, 22–24.
[17] Hardy 1991a, 30, quoting Nellie Shaw in Hardy 1979, 177.
[18] Beevers 1988, 13–14, 19, 23.
[19] Beevers 1988, 24.
[20] Beevers 1988, 24.

structural crisis. An agricultural depression, product of poor harvests and intense overseas competition following the opening-up of new land in the Americas and Australasia, reduced cereal acreage in England and Wales by one quarter between 1879 and 1900. Farm rents declined by up to 50 per cent; the Duke of Marlborough said in 1885 that if there were any effective demand, half the land of England would be on the market tomorrow; even by 1902, in Hertfordshire, it was estimated that 20 per cent of farms were unoccupied.[21]

In addition, there was a huge fallout from the "Land War" in Ireland, which in the 1880s had an impact on British politics second only to Irish Home Rule; and there was Henry George's *Progress and Poverty*, published in 1881, which sold 100 000 copies.[22] The English Land Restoration League, formed in 1883, soon became the vehicle for promoting George's ideas by "shifting all taxation on to the value of the land, irrespective of its use and improvement, and finally *taking all Ground Rent for public purposes*".[23] The idea of a single land tax gained credence through the late 1880s and early 1890s, and was embraced by a new and highly successful London evening paper, *The Star*, in 1888; at the beginning of 1889, in the first London County Council elections, taxation of land values became a popular cry, and helped secure election for radical candidates even though London was highly conservative for Parliamentary purposes.[24] Land tax was vigorously pursued in working-class organisations and, as Sidney and Beatrice Webb pointed out, it "completely revolutionised" the attitude of urban workers in the middle 1890s: "Instead of the Chartist cry of 'Back to the Land', . . . the town artisan is thinking of his claim to the unearned increment of urban land values, which he now watches, falling into the coffers of the great landlords".[25] The LCC began to urge site value rating in 1894, and was followed by many other authorities, attracting 518 in a petition of 1906; a Royal Commission on Local Taxation was split on the subject in 1901.[26]

Some who started from the Georgian tradition drifted to socialism. The Land Nationalisation Society had been set up in 1881, producing many pamphlets over many years, though the term covered a huge range from compulsory purchase to progressive nationalisation of all land for the community, a view which gained over time. Its moving spirit was Alfred Russel Wallace, an eminent scientist whose interest in land reform had been strengthened by the great Irish land debate of 1879–80. He and his fellow-spirits believed that plentiful provision of rural plots would by itself cause people to flock back to the land. He knew Howard well, and the Society was instrumental in supporting the launch of the Garden City Association in 1899, providing an initial nucleus of members, and the National Housing Reform Council in 1900.[27]

---

[21] Fishman 1977, 62.
[22] Douglas 1976, 44–45.
[23] Douglas 1976, 47, quoted in *The Christian Socialist*, July 1884, 23.
[24] Douglas 1976, 113.
[25] Douglas 1976, 117, quoting S. and B. Webb (1920), *History of Trade Unionism*, London: Longman, 376.
[26] Douglas 1976, 118–119.
[27] Aalen 1992, 45–47; Douglas 1976, 45–46, 48; Hardy 1991a, 30.

Throughout the 1880s, the land question excited huge attention. Joseph Chamberlain was the first major figure to propose land reform, in articles in the *Fortnightly Review* of 1883–85, later republished as *The Radical Programme*, although his ideas were unclear, centring on a notion of "three acres and a cow".[28] This agitation was most intense on the Celtic fringes, where it swept both Ireland and Scotland; England heard only faint echoes, for here it seemed that farm workers preferred to vote with their feet, by moving to the towns or the colonies. Yet Chamberlain made the effective point that in England they huddled in urban slums.

In particular, there was a drift of population from the heartland of British agriculture – East Anglia, and the central English counties close to London – into the capital, which grew at great speed and – coupled with conversion of homes to offices and railway building – left many people trapped in central-city slums.[29] Small wonder that the Liberals owed their success in the 1885 election partly to the appeal of Chamberlain's "three acres and a cow" to the newly enfranchised rural labourers; however, in Henry Labouchère's words, the Liberals lacked "an urban cow".[30] For they lacked an answer to the parallel issue that dominated debate in the London of the 1880s, concerning housing and living conditions in the capital.[31]

It is difficult to comprehend, now, the intense emotions the land issue raised in the last two decades of the nineteenth century. Essentially it represented a struggle for power between the old landed classes, where power still resided at the beginning of the industrial era, and new interests who wanted to dismantle the social influence of the landed estates.[32] Very often the remedies were founded on misconceptions, and were proposed by townsmen with little knowledge of rural conditions, but that was not the point: "Just as the truth of a religious doctrine is irrelevant to the mounting of a crusade, so the truth of the radical doctrines on land was irrelevant to the struggle against landed dominance so long as the potency of the symbols continued to excite the faithful".[33]

Howard was fully immersed in this debate, and in his reading he tried to develop his own solution to the problem. From Herbert Spencer he derived "the grand principle" that all men are equally entitled to the use of the earth.[34] But he did not know how to achieve it. He found his answer in a pamphlet by an obscure and eccentric radical, Thomas Spence. This pamphlet, *The Rights of Man, as Exhibited in a Lecture, Read at the Philosophical Society in Newcastle*, in November 1775, was reprinted in 1882 by H.M. Hyndman, founder of the Social Democratic Foundation, with his own notes and comments as *The Nationalisation of the Land in 1775 and 1882*; Howard must have discovered it at this point.

---

[28] Douglas 1976, 48–49.
[29] Beevers 1988, 9–10; Douglas 1976, 72, 105–106.
[30] Douglas 1976, 53.
[31] Osborn 1950, 228–229.
[32] Thompson 1965, 23–24.
[33] Thompson 1965, 24.
[34] Beevers 1988, 20.

Spence claimed "Thus were the first landlords usurpers and tyrants", as were all their successors. To remedy this, every parish should become a corporation and take up their lost rights in collectivity; rents would henceforth be paid to them, to be used for public purposes like building and repairing houses and roads. Further, these rents would soon produce a surplus for distribution to the less fortunate, and for socially improving expenditures like schools and public libraries, just as in Howard's scheme. In Spence's "Spensonia" the community would be regulated by a board of directors elected from and by the shareholders; so with the garden city.[35]

There was still a problem, for Spence nowhere explained how the people were to appropriate the land. Here Howard turned to the idea of colonisation as planned migration, which he probably first encountered in J.S. Mill's *Principles of Political Economy* around 1880. In this publication Mill called for planned colonisation along the lines advocated by Edward Gibbon Wakefield 40 years earlier, with a planned mixture of town and country. In any case, the idea of "home colonies" for the unemployed was generally current in the early 1880s, being canvassed by the Social Democratic Foundation in its pre-Marxist days, and by Keir Hardie. A leading protagonist was Thomas Davidson, an obscure Scottish–American philosopher who was one of the founders of the Fellowship of the New Life, from which the Fabian Society sprang. But Howard soon saw the problem that unemployed town workers would not easily turn to agriculture; they would need manufacturing industry.[36]

COLONISING THE LAND

He found an answer from the economist Alfred Marshall, who suggested in *The Contemporary Review* in 1884 that "there are large classes of the population of London whose removal into the country would in the long run be economically advantageous – that it would benefit alike those who moved and those who remained behind".[37] Railways, the cheap post, the telegraph, newspapers, so Marshall argued, were agents of geographical diffusion, especially for industries which followed the labour force. And this labour force was moving out: of all Londoners still living, one-fifth had left the capital.[38]

Herein, Marshall argued, was a key to action:

> the general plan would be for a committee, whether specially formed for the purpose or not, to interest themselves in the formation of a colony in some place well beyond the range of London smoke. After seeing their way to building or buying suitable

[35] Beevers 1988, 21–23.
[36] Beevers 1988, 25–26.
[37] Marshall 1884, 224.
[38] Marshall 1884, 223–225, 228.

cottages there, they would enter into communication with some of the employers of low-waged labour.[39]

"Gradually", Marshall wrote, "a prosperous industrial district would grow up; and then mere self-interest would induce employers to bring down their main workshops and even to start factories in the colony."[40]

THE CRITICAL KEYS: BELLAMY AND KROPOTKIN

By the end of the 1880s Howard had all the ideas he needed, but he still could not bring them together. The key was Edward Bellamy's *Looking Backward*, which Howard read early in 1888, shortly after its American publication. He personally testified to the influence it had on him.[41]

Bellamy was born in Chicopee Falls, Massachusetts, a small industrial town, in 1850, the same year as Howard; he died at the age of 48 in 1898, the year of publication of Howard's book. Today his book is little read by planning students. Yet in its day it sold hundreds of thousands of copies and had a profound influence on social thinking, extending for well over 40 years into the New Deal. Its central character, Julian West, takes a sleeping potion and wakes up in the Boston of 2000; in it, the nation's industrial army, working in harmony, has harnessed America's resources to a peak of efficiency, eliminating poverty, crime, greed, corruption and lack of emotional fulfilment; but it is a regimented society, in which everyone agrees with the goals of the state.[42]

This Boston of 2000 has no smoke, because of changes in energy sources; as with other nineteenth-century Utopians, Bellamy thought that technology was benign. But the industrial organisation is very Fordist, with giant integrated mills. The living conditions of the workers are improved beyond recognition.[43] Yet the city has a strangely nineteenth-century feel, for it is dense and highly developed: neither a "town in the country" nor a "country in the city". Essentially "It is a living environment similar to that of the 1890s, but which was 'regularized'".[44] It is an Haussmannesque vision, ordered and sanitised, of straight tree-lined streets, open squares full of greenery, and landscapes with fountains and statues. And, like the City Beautiful movement that was just about to burst forth in the United States, there is almost no mention of the housing of the poor.[45]

Bellamy's influence was huge but indirect, for he developed the idea of a "socialist community" which owned all the land, both agricultural and urban.[46] And this gave Howard one key element of his scheme; in 1890 he was one of the 20

---

[39] Marshall 1884, 229.
[40] Marshall 1884, 230.
[41] Beevers 1988, 26–27.
[42] Mullin and Payne 1997, 17–18.
[43] Mullin and Payne 1997, 19–20.
[44] Mullin and Payne 1997, 21.
[45] Mullin and Payne 1997, 21–25.
[46] Osborn 1946, 21.

founder-members of the Nationalisation of Labour Society, set up to promote Bellamy's ideas in England. However, the society lasted only three years, attracting the scorn of people like Sidney Webb.[47]

But soon Howard was cast again into doubt. Attracted as he might be to Bellamy's Utopia, he was fundamentally uncomfortable with its centralised socialist management and its insistence on the subordination of the individual to the group; he saw these as authoritarian.[48] And, soon after encountering Bellamy's book, he must have read the articles which Peter Kropotkin – the great Russian anarchist *émigré*, living in anomalous exile in Brighton – contributed to *The Nineteenth Century* between 1888 and 1890, which a decade later were later collected in the book *Fields, Factories, and Workshops*. They advocated the creation of "industrial villages" based on the liberating potential of electric power.[49] Later Howard called Kropotkin "the greatest democrat ever born to wealth and power" and abandoned his infatuation with Bellamy.[50]

THE NOTION OF PLANNED CITIES

Bellamy and Kropotkin, we know, were direct influences, but, somewhere along the line, Howard found other answers to the practical problems that were obsessing him, just as any obsessed inventor would. Half a century earlier, in the 1830s, Edward Gibbon Wakefield had advocated planned colonisation for the poor. The very scheme he had promoted, South Australia, provided – in Colonel Light's celebrated plan for the capital city of Adelaide – the idea that once a city had reached a certain size, planners should halt its growth by a green belt and begin a second city: the origin of the notion of Social City, as Howard acknowledged (Figures 3 and 4). James Silk Buckingham's plan for a model town, published in 1849, gave Howard the key features he used in his diagram of Garden City: the limited size (in Buckingham's case, 10 000 people), the central place, the radial avenues, the peripheral industries, the surrounding green belt and the notion of starting a further settlement once the first was full.[51] Pioneer industrial villages developed in the 1880s and 1890s in the countryside – William Hesketh Lever's Port Sunlight near Liverpool, George Cadbury's Bournville outside Birmingham – provided physical models and a practical illustration that decentralisation was indeed possible. More generally, Howard must have responded to the Back to the Land movement, which had created at least 28 more or less Utopian nineteenth-century communities, almost all rural.[52] And underlying these was the wider and looser social movement of the 1870s and 1880s, led by Morris and Ruskin, to reject

---

[47] Hardy 1991a, 31.
[48] Fishman 1977, 36; Meyerson 1961, 186.
[49] Fishman 1977, 36.
[50] Fishman 1977, 37, quoting Howard's draft of an unfinished autobiography.
[51] Ashworth 1954, 125; Benevolo 1967, 133.
[52] Darley 1975, Chapter 10; Hardy 1979, 215, 238.

Figure 3 *Colonel William Light. Adelaide's surveyor, immortalised in stone, gazes out over his concept, the basis of Howard's Social City idea.* Source: Peter Hall photograph.

Figure 4 *Adelaide and North Adelaide. Light's scheme from the air: the main city (background) and satellite city (foreground), each of limited size, separated by a permanent green belt intensively used for urban recreation.* Source: Peter Hall photograph.

industrialism in favour of a return to a rural life based on craft production and a sense of community.[53]

So all these entered the design; like any inventive tinkerer, Howard took whatever he needed, whatever seemed to work.[54] Indeed, as he recognised, there is nothing that is really new in his proposals: much earlier, Buckingham, Kropotkin, Ledoux, Owen and Pemberton all had suggested towns of limited populations with surrounding agricultural green belts; More, Saint-Simon, Fourier all had cities as elements in a regional complex.[55]

About this time – there is definite record at least from 1892 – Howard began to talk about his ideas to the more progressive London sects. He put his ideas to a wider audience early in 1893, resulting in a resolution to form a "Co-operative Land Society". The most prominent representation on his committee was the Land Nationalisation Society, with whose aims he was in broad agreement. At this point he was in favour of the municipalisation of enterprises, in effect creating a monopoly employer, though he later backed away from this. He relied on the notion of an appeal to rich people to lend money, though it later emerged that the

---

[53] Hall 1988, 91.
[54] Osborn 1950, 230.
[55] Batchelor 1969, 198.

dividend would be limited to 4 per cent.[56] By this time his "invention" was
complete in all details.[57]

A UNIQUE COMBINATION OF PROPOSALS

The ingredients, then, were nowhere original; nor did Howard ever claim they
were. What he did claim, in the famous heading of Chapter XI, was that his was "A
Unique Combination of Proposals", bringing together the proposals for organised
migration from Wakefield and Marshall, the system of land tenure derived from
Spence and Spencer, and the model city forms of Buckingham and of Wakefield as
interpreted by Light.[58] By putting them together into his "invention", Howard
believed he had solved the riddle that had vexed the land reformers for two
decades and more: how to achieve an ideal community, which could appropriate
for itself the land values it created by its own existence and its own efforts, thus
achieving land nationalisation by slow degrees, step by step, without ever
threatening or subverting the peaceful confidence of the Victorian bourgeoisie.
Indeed, bizarrely, they were to be invited to become its chief agents. Here lay the
central and startling originality of Howard's scheme. And herein lay the clue to its
success or failure.

---

[56] Beevers 1988, 34.
[57] Beevers 1988, 34–39.
[58] Howard 1898, 103.

CHAPTER 2

GARDEN CITY: IDEAL AND REALITY

Howard's scheme had two central features: its physical form and its mode of creation. Both were idealised. Both proved much harder to achieve in reality than on paper. And these difficulties had profound consequences for the subsequent history of the movement he went on to found, consequences which persist to this day.

THE PHYSICAL FORM: FROM GARDEN CITY TO SOCIAL CITY

Howard began his argument with the famous diagram of the three magnets (Figure 5). Like all the other figures, with their elaborate hand-drawn Victorian lettering – Howard drew them himself, and in the original edition they were printed in delicate pastel colours – it has a special archaic charm. But, viewed more closely, it is a brilliant encapsulation of the virtues and vices of the late Victorian English city and English countryside.

To summarise, the city had economic and social opportunity, but overcrowded housing and an appalling physical environment. The countryside offered open fields and fresh air, but there were all-too-few jobs and very little social life; and, paradoxically, if anything housing conditions for the average worker were just as bad. This contrast cannot be understood save in the context of the time: the 20 years of agricultural depression, which had brought a wave of mass migration from the countryside to the town, coupled with economic changes in London and other large cities – the huge destruction of housing for offices and railways and docks – that had crowded the new arrivals ever more densely into slum tenements. Howard wrote: "It is well-nigh universally agreed by men of all parties, not only in England, but all over Europe and America and our colonies, that it is deeply to be deplored that the people should continue to stream into the already over-crowded cities, and should thus further deplete the country districts."[1] In support he quoted Lord Salisbury, sometime Chair of the London County Council, Sir John Gorst and Dean Farrar, as well as several newspapers.[2] The problem, then, was to reverse the flow of migration.

---

[1] Howard 1898, 2–3.
[2] Howard 1898, 3–4.

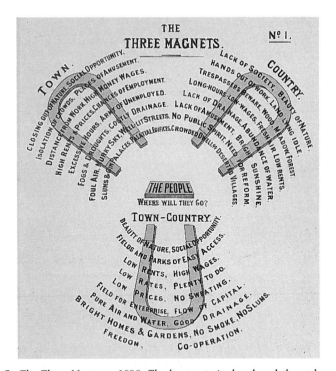

Figure 5 *The Three Magnets, 1898. The language is dated and the style archaic, but the message is still startlingly apposite; compare Figure 34.* Source: Howard 1898.

Yes, the key to the problem how to restore the people to the land – that beautiful land of ours, with its canopy of sky, the air that blows upon it, the sun that warms it, the rain and dew that moisten it – the very embodiment of Divine love for man – is indeed a *Master-Key*, for it is the key to a portal through which, even when scarce ajar, will be seen to pour a flood of light on the problems of intemperance, of excessive toil, of restless anxiety, of grinding poverty – the true limits of Government interference, ay, and even the relations of man to the Supreme Power.[3]

The answer was to ask precisely what was this quality of *magnetism*, that drew people to the city: "Each city may be regarded as a magnet, each person as a needle; and, so viewed, it is at once seen that nothing short of the discovery of a method for constructing magnets of yet greater power than our cities possess can be effective for re-distributing the population in a spontaneous and healthy manner."[4] And this magnet consisted in opportunity: "The town is the symbol of society – of mutual health and friendly co-operation, of fatherhood, motherhood, brotherhood, sisterhood, of wide relations between man and man – of broad, expanding sympathies – of science, art, culture, religion."[5] The country could not

---

[3] Howard 1898, 5.
[4] Howard 1898, 6.
[5] Howard 1898, 9.

offer those advantages – but it had something else: "The country is the symbol of God's love and care for man. Our bodies are formed of it; to it they return."[6]

So he asked himself the rhetorical question:

> "What," some may be disposed to ask, "can possibly be done to make the country more attractive to a work-a-day people than the town – to make wages, or at least the standard of physical comfort, higher in the country than in the town; to secure in the country equal possibilities of social intercourse, and to make the prospects for the advancement of the average man or woman equal, not to say superior, to those enjoyed in our large cities?"[7]

The clue was that it was possible to create yet a third form of living and way of life, superior to either.

> There are in reality not only, as is so constantly assumed, two alternatives – town life and country life – but a third alternative, in which all the advantages of the most active and energetic town life, with all the beauty and delight of the country, may be secured in perfect combination; and the certainty of being able to live this life will be the magnet which will produce the effect for which we are all striving – the spontaneous movement of the people from our crowded cities to the bosom of our kindly mother earth, at once the source of life, of happiness, of wealth, and of power.[8]

In this way, by creating the third magnet, it would be possible to square the circle: to gain all the opportunities of the town, all the qualities of the country, without any degree of sacrifice: "Town and country *must be married*, and out of this joyous union will spring a new hope, a new life, a new civilisation."[9] He promised that

> I will undertake, then, to show how in "Town–country" equal, nay better, opportunities of social intercourse may be enjoyed than are enjoyed in any crowded city, while yet the beauties of nature may encompass and enfold each dweller therein; how higher wages are compatible with reduced rents and rates; how abundant opportunities for employment and bright prospects of advancement may be secured for all; how capital may be attracted and wealth created; how the most admirable sanitary conditions may be ensured; how excessive rainfall, the despair of the farmer, may be used to generate electric light and to propel machinery; how the air may be kept clear of smoke; how beautiful homes and gardens may be seen, on every hand; how the bounds of freedom may be widened, and yet all the best results of concert and co-operation gathered in by a happy people.[10]

The way to achieve this, Howard argued, was to create a totally new town in the middle of the countryside, outside the sphere of the big city, where land could be

---

6   Howard 1898, 9.
7   Howard 1898, 6.
8   Howard 1898, 7.
9   Howard 1898, 10.
10   Howard 1898, 10.

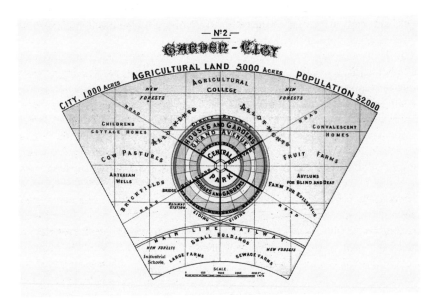

Figure 6 *Garden City. The basic notion: a mixed-use, medium-density, fixed-size development; jobs, schools, shops, parks, countryside all within walking distance.* Source: Howard 1898.

bought at depressed agricultural land values. This *Garden City* would have a fixed upper limit – Howard suggested 32 000 people, living on 1000 acres (405 ha) of land, about one and a half times the size of the historic City of London. It would be surrounded by a much larger area of permanent green belt, bought and owned by the garden city management as part of the purchase package – Howard proposed 5000 acres, or 2023 ha – containing not merely farms, but also all kinds of quasi-urban institutions, like reformatories and convalescent homes, that could benefit from a rural location (Figure 6).

Garden city would be a true small-to-medium-sized town, offering the usual range of urban jobs and services. In 1898, jobs meant factory jobs. He illustrated them in painstaking detail: clothing, cycles, engineering, jam-making. They were what came to be known as light industries, since – as Howard himself emphasised – the industries attracted out here would be the ones where the quality of the workforce would be the prime concern. Howard believed that industrialists would gladly follow the lead already set by pioneers like Cadbury at Bournville and Lever at Port Sunlight; they would see the advantages of operating in a clean smoke-free atmosphere where their workers would be healthier and closer to their work than in the giant city.

Their homes would be close by, because the entire Garden City would be circular in form, with a radius of only three quarters of a mile (1.2 km) from centre to edge. It would be traversed by six wide radial boulevards, dividing it into six equal sections or wards. (Thirty years later, the American sociologist–planner

Clarence Perry would re-invent the concept under the label of "neighbourhood units", a term that became incorporated into British planning practice after World War II.)

In the very centre of the town, instead of the usual dense collection of offices and shops, there would be a public garden 5 acres (2 ha) in size, bigger than Trafalgar Square, and this would be surrounded by a ring of impressive public buildings: the town hall, a concert hall, a lecture hall, museum, picture gallery, library and theatre. But these buildings would also look outwards on a much larger central park, no less than 150 acres (60 ha) in size, or roughly equal to the size of Hyde Park and Kensington Gardens, with "abundant provision for football, cricket, tennis and other outdoor games".[11] The inspiration here might possibly have been the centre of a city like Washington, DC, which Daniel Burnham was just about to restore to its original glory: the Congress and the White House and other great public buildings, set off against monumental open space. Or perhaps Howard took his inspiration from closer to home, from the Horse Guards and Buckingham Palace set against the green space of St James's Park. All this would be possible because Howard was not constrained by traditionally high urban land values; he was free to put a park in the centre, just as Henry VIII had been free over three centuries before, for somewhat similar reasons.

But then Howard added a startling feature which was quite his own. He describes it in an article submitted to the *Contemporary Review*, which was rejected but which survives in his unpublished papers:

> Running all around the Central Park is a wide glass Arcade or Crystal Palace. This building is in wet weather one of the favourite resorts of the people; for the knowledge that its bright shelter is close at hand will tempt people into the park even in the most doubtful of weathers. Here manufactured goods are exposed for sale, and here most of the shopping which requires the job of deliberation and selection is done. The space is however a good deal larger than is required for these purposes, and a considerable part of it is used as a winter garden, and the whole forms a permanent exhibition of a most attractive character – the furthest inhabitant being within 600 yards.[12]

It was startling for several reasons. First was the precision whereby Howard ensured that the shopping centre would be available to every inhabitant within convenient walking distance. But second was the inspiration: the Crystal Palace was clearly at one level an imitation of the arcades which he must have seen in the West End of London (and maybe in provincial cities such as Leeds, where they are a prominent feature to this day); at another, it specifically derives from the Winter Gardens which were then just becoming a prominent feature of English seaside resorts; at a third, there was a direct link to Paxton's Crystal Palace, which Howard must have visited at Sydenham, close to his first marital home,[13] and perhaps to Alexandra Palace, near his later Stamford Hill home. But the greatest irony is that the Crystal Palace now appears startlingly modern, for it is clearly the direct

---

[11] Quoted in Beevers 1988, 50.
[12] Quoted in Beevers 1988, 52.
[13] Beevers 1988, 53.

precursor of all the great enclosed shopping malls, born in the United States in the 1950s, which now crown our city centres and new edge-of-town centres.

On the outer side of Crystal Palace began the residential area, forming an outer ring some 750 feet (230 metres) wide. The houses would be arrayed along the boulevards and the intermediate radials – two of them between each pair of radials – and along five circumferential avenues that linked them. (Howard's American experience clearly shows in the differentiation between streets and avenues and in the numbering of the avenues, both typical of New York; though the plan is again more reminiscent of Washington.) But then there was another feature: the third avenue, or Grand Avenue.

> This Avenue which forms a belt of green two and three-quarter miles long divides that part of the town which lies outside the Central Park into two belts of equal width, really constitutes an additional park which is within 240 yards of the farthest removed inhabitant of the town. It is of the remarkable width of 420 feet and represents, exclusive of the roads which intersect it, 118 acres. In this splendid Avenue, sites (24 acres in extent) are reserved for schools and their surrounding playgrounds, while other sites are reserved for churches.[14]

The inspiration again appears American: the Mall in Washington, for instance, but most likely is Chicago's Midway, a similar wide strip that Daniel Burnham designed for the 1893 Columbian Exposition, which now divides the Hyde Park district on the South Side of Chicago.[15]

Howard's vision of the residential areas would have looked quite different from their actual realisation in the first Garden City of Letchworth a few years later, which we owe to the architects Raymond Unwin and Barry Parker. For there was to be no single architectural style; on the contrary, they were to be "of the most varied architecture and design that ingenuity and individuality can suggest – a general observance of street line or harmonious departure from it being the chief point over which the Board of Management exercises control".[16] Here, again, the image is of a superior Victorian suburb of the time, whether in Britain or America; Chicago's Hyde Park again springs to mind.

But there was one important difference: density. As Lewis Mumford first pointed out in his celebrated introduction to the re-issue of the second edition of Howard's book, in 1946, Howard's assumptions about density were "on the conservative side; in fact, they followed the traditional dimensions that had been handed down since the Middle Ages, and, one may add by way of criticism, followed them too closely".[17] Howard suggested that the average lot should be 20 by 130 feet (c. 6 × 40 m), and the minimum 20 by 100 (c. 6 × 30 m). Such lot sizes were completely consistent with traditional cities of the time: 20 by 100 is, for instance, the typical New York City lot. With five people to a family, typical in 1898, this gave a density of about 90–95 persons per acre (222–235/ha); with the smaller family units of

---

[14]   Quoted in Beevers 1988, 52.
[15]   Stern 1986, 309; Girouaud 1985, 317.
[16]   Quoted in Beevers 1988, 52.
[17]   Mumford 1946, 31–32.

1946, it would give about 70 per acre (173/ha)[18] and with the even smaller units of the 1990s, perhaps half that. It is remarkable that when Abercrombie and Forshaw developed their reconstruction plans for inner London, in 1943, their proposed density of 136 per acre (336/ha) – little above Howard's – required eight- and ten-storey tower blocks; but Howard based his plan on conventional single-family homes, with gardens, throughout his garden city. Mumford commented: "This twenty-foot front is far too narrow for a good modern building row, with relatively shallow rooms, fully open to the penetration of the sun's rays."[19]

Finally, there were the jobs. To put them as close as possible to the homes of the workers, Howard provided for workplaces in a narrow industrial belt around the edge of the town, served by a circular railway; he was writing two years after the repeal of the law that said a man with a red flag had to walk in front of any motor vehicle, so he can perhaps be excused for failing to predict the impact of the truck and the van on industrial location. If we substitute a circular highway linked to a motorway, it again looks startlingly modern – though perhaps less sustainable.

And sustainability, to use the overworked 1990s term, was what Garden City was all about. The astonishing fact about Howard's plan is how faithfully it follows the precepts of good planning a century later: this is a walking-scale settlement, within which no one needs a car to go anywhere; the densities are high by modern standards, thus economising on land; and yet the entire settlement is suffused by open space both within and outside, thus sustaining a natural habitat.

Nowhere is this principle better seen, than in Howard's treatment of the Garden City's growth. As more and more people moved from the congested metropolis into his Garden City, it would before long reach its planned limit of 32 000 people; then, another city would be started a short distance away; then another, then another. The result, over the course of time, would not be a single Garden City but an entire cluster of such places, each Garden City offering a range of jobs and services, but each connected to all the others by a rapid transit system (or, as Howard quaintly called it, an Inter-Municipal Railway), thus producing all the economic and social opportunities of the giant city. Howard called this polycentric vision Social City (Figure 7). It was, he wrote,

> a carefully-planned cluster of towns, so designed that each dweller in a town of comparatively small population is afforded, by a well-devised system of railways, waterways, and roads, the enjoyment of easy, rapid, and cheap communication with a large aggregate of the population, so that the advantages which a large city presents in the higher forms of corporate life may be within the reach of all, and yet each citizen of what is destined to be the most beautiful city in the world may dwell in a region of pure air and be within a very few minutes' walk of the country.[20]

In the diagram which Howard appended in the 1898 edition, but which was sadly omitted afterwards, Social City covered 66 000 acres (26 700 ha), slightly less than the area of the old London County Council of Howard's day; it had a total

---

[18] Mumford 1946, 32.
[19] Mumford 1946, 32.
[20] Howard 1898, 131.

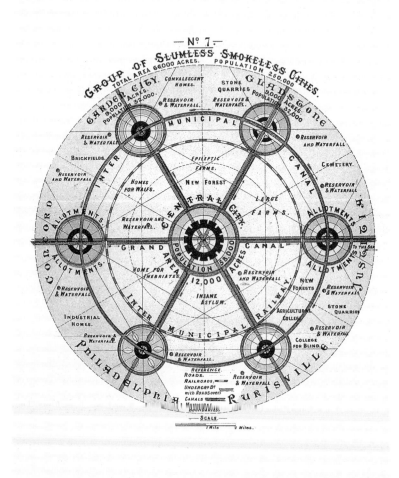

Figure 7 *Social City. The key to the puzzle: clusters of Garden Cities, each "sustainable" by the standards of the 1990s, linked by a rapid transit system; an amazingly modern concept.* Source: Howard 1898.

population of a quarter of a million, equal to a major English provincial town like Hull or Nottingham at that time.[21] In fact, it is quite clear that Social City could proliferate almost without limit, until it became the basic settlement form covering most of the country. Because the diagram was replaced by a truncated version in all subsequent editions, most readers have failed to grasp the vital fact that Social City, not the individual isolated Garden City, was to be the physical realisation of Howard's third magnet.

THE FINANCIAL KEY: THE CAPTURE OF DEVELOPMENT VALUE

That, then, was the physical expression. But equally novel was the financial model by which Howard proposed to create it. The key was that the land for each Garden City and its surrounding green belt, an area of 6000 acres (2400 ha), would be purchased in the open market at depressed agricultural land values: £40 an acre (£100/ha), or £240 000 in all, the money to be raised on mortgage debentures paying 4 per cent.[22] This land would be legally vested in four gentlemen "of responsible position and undoubted probity and honour, who hold it in trust, first, as a security for the debenture-holders, and, secondly, in trust for the people of Garden City".[23]

Over quite a short period of time, Howard then argued, the growth of Garden City would begin to raise land values, and thus the rents that would be payable. At the time, £4 an acre was a very high rent for agricultural land, yet land in the centre of London would be worth £30 000 an acre. Howard wrote:

> The presence of a considerable population thus giving a greatly added value to the soil, it is obvious that a migration of population on any considerable scale to any particular area will be certainly attended with a corresponding rise in the value of land so settled upon, and it is also obvious that such increment of value may, with some foresight and pre-arrangement, become the property of the migrating people.[24]

Specifically, the entire basis of the financing was that rents could and would be regularly revised upwards, in line with the general rise in land values; and this would allow the four "responsible gentlemen" not only to pay off the mortgage debt, but increasingly to generate a fund for social purposes. Howard explained the novelty of his scheme:

> Amongst the essential differences between Garden City and other municipalities, one of the chief is its method of raising its revenue. Its entire revenue is derived from rents; and one of the purposes of this work is to show that the rents which may very

---

[21] Howard 1898, 131.
[22] Howard 1898, 12.
[23] Howard 1898, 13.
[24] Howard 1898, 21.

reasonably be expected from the various tenants on the estate will be amply sufficient, if paid into the coffers of Garden City, (a) to pay the interest on the money with which the estate is purchased, (b) to provide a sinking-fund for the purpose of paying off the principal, (c) to construct and maintain all such works as are usually constructed and maintained by municipal and other local authorities out of rates compulsorily levied, and (d) (after redemption of debentures) to provide a large surplus for other purposes, such as old-age pensions or insurance against accident and sickness.[25]

Howard divides the "rent" payable into three parts: one representing interest on the debentures, called "landlord's rent", another repayment of purchase money, the "sinking-fund", and a third devoted to public purposes, "rates", the whole being termed "rate-rent".[26] He shows in a diagram – The Vanishing Point of Landlord's Rent – how, as the first two are gradually paid off, the entire rent yield is progressively applied to the creation of a local welfare state, entirely without need for local or central taxation, and directly responsible to the local citizens (Figure 8). The administration of Garden City would be able "to found pensions with liberty for our aged poor, now imprisoned in workhouses; to banish despair and awaken hope in the breasts of those that have fallen; to silence the harsh voice of anger, and awaken the soft notes of brotherliness and goodwill".[27]

Today's reader may be surprised that so much of the book consists of financial calculations. The reason, simply, is that Howard was directing it to hard-nosed Victorian businessmen who would have to be assured that their money would be safe in the hands of the four gentlemen, however respectable they might be. This was not as completely idealistic as it might seem. In the low-inflation climate of Victorian England, stock like consols might pay as little as 2 per cent a year; businessmen with a social conscience were already quite accustomed to the notion of philanthropy and 5 per cent, so maybe they could be persuaded to accept a little less.

Social City was thus not merely a sustainable physical form; it was a brilliant marketing device. As Howard saw it, the first Garden City would act as a shining exercise in public relations, giving clear evidence to a nervous investing public that the entire idea was sound and was gaining momentum. Social City would be achieved incrementally; the greater the success, the easier the money would be to raise.

He even thought that the construction of social cities would cause ground rents to fall in London, though out-migration would cause the fixed burden of rates to rise. Rents in London would fall, and rents of slum property would fall to zero, so that it could be torn down and replaced by parks, gardens and allotments. Howard laid down that the area of the London County Council should contain at most one fifth of its then population, or about 800 000;[28] in 1998, having lost about half its population, it still has more than double that.

25   Howard 1898, 20–21.
26   Howard 1898, 28.
27   Howard 1898, 141.
28   Howard 1898, 145–150.

Figure 8 *The Vanishing Point of Landlord's Rent. The neglected Howard diagram, showing the key point in his calculations: by buying land at depressed agricultural prices, the community could create its own urban land values and use them to finance a local welfare state.* Source: Howard 1898.

Howard was at pains to distinguish his scheme from "socialism", for he felt that socialism ignored the individualism and self-seeking quality of human beings.[29] Against Hyndman and other socialists he argued that "in striving to become possessed of present wealth forms, he is laying siege to the wrong fortress".[30] For instead of seeking to acquire all existing land, it would be possible to proceed one step at a time.[31] The important point about Howard's scheme was that it was voluntary: "my proposal appeals not only to individuals but to co-operators, manufacturers, philanthropic societies, and others experienced in organisation, and with organisations under their control, to come and place themselves under conditions involving no new restraints but rather securing wider freedom".[32] For Howard, Garden City was far more than just a town: it was a third socio-economic system, superior both to Victorian capitalism and to bureaucratic centralised socialism.

THE PHILOSOPHY: ANARCHISM AND CO-OPERATION

The words at the bottom of the Three Magnets diagram, FREEDOM – CO-OPERATION, were thus no mere rhetorical flourish. For each Garden City would be an exercise in local management and self-government. Services would be provided by the municipality, or by private contractors, as proved more efficient. Others would come from citizens, in a series of what Howard called pro-municipal experiments – or self-help. In particular, people would build their own homes with capital provided through building societies, friendly societies, co-operative societies, or trade unions.

It was a vision of anarchist co-operation, to be achieved without large-scale central state intervention. Not for nothing did Howard admire Kropotkin. Garden City would be realised through individual enterprise, wherein individualism and co-operation would be happily married. It would employ "the very highest talents of engineers of all kinds, of architects, artists, medical men, experts in sanitation, landscape gardeners, agricultural experts, surveyors, builders, manufacturers, merchants and financiers, organizers of trade unions, friendly and co-operative societies, as well as the very simplest forms of unskilled labour, together with all those forms of lesser skill and talent which lie between".[33]

But here the first doubt enters in. There was to be a "Board of Management" consisting of "The Central Council", in which were vested the rights and powers of the community as sole landlord, and which was to receive all the rate–rents after deduction of landlord's rent and sinking-fund; and "The Departments" to which were delegated specific functions – Public Control (finance, assessment, law,

[29] Howard 1898, 97–98.
[30] Howard 1898, 121.
[31] Howard 1898, 123.
[32] Howard 1898, 100.
[33] Howard 1898, 140.

inspection), Engineering, and Social and Educational.[34] But, apparently quite
separately, there would be the limited-dividend company, concerned with the
return on the invested capital. Howard did not seem seriously to have grasped that
the two bodies might come into conflict. But this would emerge all too soon.

FROM IDEA TO ACTION: THE GARDEN CITY ASSOCIATION

Howard's book was respectfully received, but it did not make huge waves. As F.J.
Osborn commented half a century later, "no book of significance has enjoyed less
academic notice or prestige. With the exception of Alfred Marshall and Charles
Gide no economist of the first rank has, until the last few years, seriously
entertained Howard's central idea that the size of towns is a proper subject of
conscious control."[35]

There was good reason for this academic neglect, Osborn thought:

> Howard did not seem a "scientific" writer; his book avoids technical terminology,
> displays no great learning, contains little historical or demographic documentation.
> Neither did it become a best-seller exercising that massive influence on popular
> attitudes which students of social affairs are reluctantly compelled to respect. Yet it is
> surprising to me that so few trained thinkers detected that Howard possessed
> extraordinary intuition and judgment; that he had descried and fastened upon a
> neglected problem of major importance; and that in his chosen field he had a shrewd
> instinct for discriminating the permanently significant from the ephemeral in the
> ideas of his time . . . his assumptions were in fact almost wholly right, because they
> were based on a wide sympathy with the habits and desires of common people.[36]

But Howard was not to be deterred. A mere eight months later – on 21 June
1899, at a meeting at the Memorial Hall in Farringdon Street, London – Howard
took the lead in setting up a Garden City Association.[37] Its objectives were:

> To promote discussion of the project as suggested by Mr Ebenezer Howard in his book
> *To-morrow*.
> To take the initial steps towards the formation in Great Britain, either by public
> Company or otherwise, of Garden Cities, wherein shall be found the maximum
> attainable of comfort and convenience to the inhabitants, who shall themselves
> become in a corporate capacity the owners of the site subject to the fullest
> recognition of individual as well as mutual and public interests.[38]

Howard took care to make the new Association politically bipartisan and to
harness the energies of manufacturers, merchants and financiers as well as co-
operators, artists and ministers. By 1902, when his book came out in a second

---

[34] Howard 1898, 67, 70.
[35] Osborn 1946, 9–10.
[36] Osborn 1946, 10.
[37] Beevers 1988, 72.
[38] Hardy 1991a, 42.

edition, membership had risen to over 1300, of whom 101 were Vice-Presidents; there were two peers, three bishops and 23 Members of Parliament, a few academics including Marshall, and half a dozen industrialists, including Cadbury, Lever and Rowntree. Ralph Neville, a distinguished lawyer who soon afterwards became a judge, was passionately devoted to the cause and accepted the offer of Chairmanship of the GCA Council in 1901, injecting some hard practical sense; and he was fortified by the appointment of a young surveyor, Thomas Adams, as paid secretary. But Howard was less than successful in winning over the support of some groups of people: there was very little business expertise, or representation of the Labour movement.[39]

The Fabian Society, some of whose leading lights were on close terms with Howard, were openly contemptuous. And with good reason: his book flatly contradicted their basic strategy, developed by Sidney Webb, of achieving socialism through municipal ownership within the existing cities, or Gas and Water Socialism. Edward Pease, their secretary, ridiculed Garden Cities as such: "The author has read many learned and interesting writers, and the extracts he makes from their books are like plums in the unpalatable dough of his utopian thinking."[40] Howard persisted and even got an invitation to speak to the Society, when he carefully tailored his message to them. Some prominent Fabians, such as H.G. Wells, joined the Garden City Association, but the Society continued to cold-shoulder him.[41]

The most prominent Fabian of all, George Bernard Shaw, had a "strangely ambivalent" relationship to the movement, which would continue for 25 years, almost to his death.[42] In a letter he describes a lecture by Howard:

> We went down to Hindhead from Saturday to yesterday. On Monday Ebenezer the Garden City Geyser lectured in Hindhead Hall, with a magic lantern giving views of that flourishing settlement in the manner of Mr. Scadder in Martin Chuzzlewit. I had to make a speech which had so fell an effect, in spite of my earnest endeavors to help him over the stile, that the audience declined to put up a single hand for the resolution. Finally the chairman put it again, coupling it with a vote of thanks, when, the situation becoming too poignant, I ostentatiously held up my paw, on which the others followed suit and Eb was saved. I pointed out that manufacturers were ready enough to go into the country; but what they went there for was cheap labor. I suggested that half a dozen big manufacturers building a city could give good wages, and yet get so much of them back in rent and shop rent, or in direct butcher, baker and dairy profits, that the enterprise might pay them all the same. At this the Hindhead proletariat grinned from ear to ear, and concluded that I was the man who really understood the manufacturing nature, the Geyser being a mere spring of benevolent mud.[43]

Here, Shaw proved highly prescient, as events were soon to show.

---

[39] Beevers 1988, 72–73, 79–80; Macfadyen 1933, 37.
[40] Beevers 1988, 71.
[41] Beevers 1988, 71.
[42] Beevers 1988, 69–70.
[43] Beevers 1988, 70.

Howard was equally unsuccessful, much more surprisingly, with the Co-operative movement. For, right from the beginnings of them around 1800, the stores were not merely agencies for bulk purchase of goods but were also outlets for the sale of goods made in co-operative workshops.[44] And the original rules of the pioneer Rochdale Society, in 1844, made provision for

> The establishment of a store for the sale of provisions, clothing, etc.
> The building, purchasing or erecting a number of houses, in which those members desiring to assist each other in improving their domestic and social conditions may reside.
> To commence the manufacture of such articles as the society may determine upon; for the employment of such members as may be without employment, or who may be suffering in consequences of repeated reductions in their wages.
> . . . the society shall purchase or rent an estate or estate of land, which may shall be cultivated by the members who may be out of employment, or whose labour may be badly remunerated.
> That as soon as practical, this society shall proceed to arrange the powers of production, distribution, education, and government; or in other words, to establish a self-supporting home colony of united interests, or assist other societies in establishing such colonies.[45]

It could not be more clear. But Beatrice Webb showed, in her early study of co-operation, that as the movement grew, it did so through a system of consumers' co-operatives, not a producers' one.[46] Despite that, Howard and his supporters hoped that the co-operative movement would be the main builders of Garden City, and at each Co-operative Congress from 1900 to 1909 they argued that the movement's stores, factories and homes should be concentrated in the first Garden City then under construction in Letchworth. But despite influential supporters among national leaders, the individual distributive societies were too concerned with their independence.[47]

The Land Nationalisation Society continued to support Howard, as it had from the start; it provided office space to the new association and also its first secretary in the person of F.W. Steere, a barrister; many prominent members were attracted to Howard's ideas because they seemed a non-alarming and thus practicable way of securing progressive land nationalisation.[48]

Howard and his band of supporters pressed on regardless. In July 1902, the Garden City Pioneer Company was registered with a capital of £20 000, in order to survey potential sites.[49] The directors laid down criteria closely following Howard's: a site of between 4000 and 6000 acres (1620 and 2430 ha), with good rail connections, a satisfactory water supply and good drainage. The favourite site, Childley Castle east of Stafford, was rejected as too far from London.

---

[44] Bailey 1955, 12.
[45] Bailey 1955, 19–20.
[46] Webb 1938, 431–432.
[47] Fishman 1977, 65.
[48] Beevers 1988, 71–72.
[49] Macfadyen 1933, 37–39; Simpson 1985, 14.

Letchworth, 35 miles (56 km) from London in an area of severely depressed agriculture and low land prices, met the criteria and – after delicate secret negotiations with 15 owners – the 3818-acre (1545 ha) site was bought for £155 587. The First Garden City Company was registered on 1 September 1903, with a £300 000 capital, of which £80 000 was to be raised immediately, and a 5 per cent dividend.[50]

As the GCA moved into the hard job of actually building a new Garden City, a stronger business presence was injected into the movement. The Garden City Pioneer Company had Neville in the Chair; its members included Edward Cadbury, T.H.W. Idris, Howard Pearsall (a civil engineer), Franklin Thomasson (a cotton spinner), Thomas Purvis Ritzema (a newspaper proprietor) and Aneurin Williams (an ironmaster), plus Howard. They had good press contacts through Alfred Harmsworth who subscribed £1000 and offered free advertising space in the *Daily Mail*. The First Garden City Company had the same board, with the addition of two members including the soap magnate William Hesketh Lever. Raymond Unwin and Barry Parker were hired as architect–planners, beginning work at the start of 1905; Unwin (Figure 9) treated Howard's diagram with respect, modifying it to suit a site that was bisected by a railway line, but also bringing to the task his own passionate belief in the qualities of the small medieval town. There was very successful publicity, and in the summer of 1905 some 60 000 people came to see progress with the new city.[51] But, as Dennis Hardy concludes,

> The kind of money that was needed to build a new city drew Howard away from any immediate hope of financing the venture primarily from within the ranks of fellow radicals, excited by the prospect of a "co-operative commonwealth", and, increasingly, into the world of Edwardian company boardrooms and the panelled lounges of gentlemen's clubs.[52]

As Robert Fishman has commented, instead of a peaceful alternative to capitalism the Garden City became a device for preserving it.[53]

And now came the problem that Shaw, with his usual brilliance, had grasped so quickly and so clearly. After the GCA's Bournville conference of 1901, Shaw wrote a 20-page letter to Ralph Neville; there is no evidence it was sent, but it seems very likely that it was. In it Shaw affirmed his support for the idea, but questioned whether capitalists would ever agree to a trust deed limiting their freedom; he claimed that they would want to keep control of the enterprise, and that they would have the upper hand; they might provide better than average housing, but they could not be controlled; they would not distribute their profits, though they might tolerate a maximum 5 per cent dividend on the shares in the city trust; they would decline all obligations as to trade unionism, temperance, co-operation, moral reform or anything like that. The only way out

[50]   Culpin 1913, 16; Simpson 1985, 14–17.
[51]   Beevers 1988, 86, 98–100; Hardy 1991a, 47, 52.
[52]   Hardy 1991a, 47.
[53]   Hardy 1991a, 47.

Figure 9 *Raymond Unwin. With his partner Barry Parker, Unwin achieved the perfect physical realisation of the Garden City in a style influenced by Ruskin, Morris and the Arts and Crafts movement.* Source: Town and Country Planning Association.

of this dilemma might be to nationalise the Garden City like the telegraphs and public roads.[54]

Neville must have realised that Shaw's argument completely negated the basic idea that the freehold must be and must remain forever inalienable and the unearned increment secured for the community.[55] But Shaw was soon proved right. From the outset Letchworth was chronically undercapitalised: at the formal opening, 9 October 1903, only £40 000 out of a projected £300 000 had been subscribed, all by the Directors – barely a quarter of the purchase cost of the land. Some £60 000 of shares were sold to the public in the first year, but it took three years to reach £150 000; meanwhile, over those three years the company was forced to spend over £600 000 to provide the roads, gasworks, electrical generators and other utilities the town needed; the company was unwilling to borrow money for these, but was also unwilling to go into debt. For a long time it was not possible to build houses, shops, factories or public buildings. A dividend to the stockholders was not forthcoming until 1913, and then only at 1 per cent; only in 1923, after 20 years, was the full dividend of 5 per cent declared; and payment of arrears on the limited dividend was not completed until 1945.[56]

Partly as a result, from the very beginning of the Letchworth enterprise, "power was to reside in the hands of a small group of directors responsible to their shareholders for achieving a relatively narrow set of objectives".[57] Howard was appointed salaried Managing Director, and was now free to devote himself entirely to his cause; at 53, he was at the height of his powers. But he clearly was not very successful in his new role, and a breach took place with his fellow-directors when, on 1 September 1903, First Garden City Ltd was set up. He demonstrated a lack of realism as to how the money was to be raised, apparently thinking that subscribers would not be concerned about the level of dividend or when it would be forthcoming. By mid-1903 Neville was politely lecturing him about his public views that people should invest from philanthropic motives. By November the Directors had carefully acted to exclude him from any managerial function; maybe he already realised that he was not cut out for the work.[58]

One major difference was in self-government, where Howard soon found out that Neville and his fellow-directors had no intention of alienating potential shareholders. The fact was that in his book he had failed properly to address the problem of the relationship between the trustees and the community; he had become aware of this soon after foundation of the GCA, but when the First Garden City Company was created there was no mention in its Memorandum and Articles of Association of any legal obligation to transfer power progressively to the community. And, though Howard persisted in asserting that this was the ultimate intention of the Board, the Company's solicitor advised against it, and the idea was quietly dropped.[59] Thus

---

[54] Beevers 1988, 73–76.
[55] Beevers 1988, 77.
[56] Creese 1966, 215–216; Fishman 1977, 71.
[57] Beevers 1988, 82.
[58] Beevers 1988, 82, 86–89.
[59] Beevers 1988, 90.

"His influence on policy, both in its long-term implications and its day-to-day application, steadily diminished; as a direct result vital features of Howard's concept of the garden city were successively discarded."[60]

By 1904 it was clear that firms were not being attracted; the Board had neither factories nor housing to offer, only sites. In the first two years, only about a thousand people came. Industry proved difficult to attract; it was a major breakthrough when the printing and binding works of J.M. Dent, a major publisher, was attracted.[61] So the first inhabitants were the idealist, artistic middle class, who gave Letchworth a permanent reputation for crankiness that it later ill-deserved:

> here was a whole colony of eccentrics making an exhibition of themselves rather too near our sacred borders. We wished they would remove their mad city a little nearer Arlesley.[62]

Arlesley was the local mental institution. That was doubtless overdone, but there were grounds for suspicion.[63]

The Board began to falter; Lever, who wanted to offer freehold sites, resigned. A major debate followed on "Point 4" of the Association's objectives, on the gain in value to the community; Neville said that it was not realistic in this first attempt, and the Directors agreed that the "Improvement Rate" was a serious disincentive. The result was a compromise: tenants would be offered a choice between a "Howard lease" with ten- (not five-) year revisions and a normal 99-year fixed lease; most, predictably, insisted on the second, which became typical.[64] "Howard, who was still earning his living as a stenographer, was no match for a cocoa millionaire or a soap magnate", and took the defeat in good spirits, believing that his system would come to operate in the fullness of time.[65] But ironically, by their victory these sound businessmen had secured the failure of the scheme: "The unearned increment was there but without any mechanism for translating it into ready cash."[66]

The Directors proved equally conservative on other matters, and when in 1905 they gave away part of the proposed town centre – a Buxton-style Crescent – for a cottage housing exhibition that was to become permanent, Raymond Unwin had nothing more to do with it, effectively abandoning Letchworth for Hampstead Garden Suburb. In fact even though the cottage housing exhibition showed that houses could be built for as little as £150 and let at rents from 4s.6d. (22p) to 16s.6d. (82p) per week, even this was too much for unskilled workers who had to find poorer housing outside the Garden City. Many of the first houses were built

---

[60] Beevers 1988, 91.
[61] Jackson, F. 1985, 71; Simpson 1985, 20, 35.
[62] Macfadyen 1933, 47.
[63] Marsh 1982, 238–239.
[64] Beevers 1988, 93–96.
[65] Beevers 1988, 96; Fishman 1977, 66.
[66] Creese 1966, 216.

by speculative contractors whose designs introduced eccentricities that Parker and Unwin had wanted to banish.[67]

But Unwin and Parker had indelibly left their mark – to the degree that ever after, Howard's Garden City has been associated with the physical expression they gave to it (Figures 10 and 11). It is unclear, and perhaps will remain for ever unclear, how far they subtly traduced his original vision. And perhaps the question is irrelevant: the vision was large enough to contain many different expressions, as the contrast between Letchworth and the second Garden City at Welwyn immediately demonstrates. The fact is that their philosophies were different: Unwin and Parker lacked Howard's confidence in industrialisation, harking back to a pre-industrial world and idealising the traditional English village; they were echoed by Liberals like Cadbury, who looked back to an imagined paternalistic order.[68] Lewis Mumford long after commented that

> When Messrs. Unwin and Parker came to design Letchworth itself, they perhaps leaned over backwards, in their effort to avoid mechanical stereotypes, in order not to duplicate Howard's diagrammatic city. Unwin's love for the rambling layout of medieval German hill towns was even in some degree at war with Howard's rational clarifications and forward-looking proposals.[69]

Unwin and Parker wanted co-operative living in quadrangles, but only one such experiment, Homesgarth at Letchworth, was actually built; Howard supported it for a typically pragmatic reason – he thought it would appeal to the hard-up middle class who needed domestic help.[70]

We will never know whether these bolder experiments might have worked, had they been repeated more insistently. Howard liked Letchworth enough to live there for a while, but then he liked Welwyn well enough too. Perhaps the difference between neo-German-vernacular and neo-English-Georgian-vernacular was something that passed him by. He would have approved of what his ordinary neighbours approved; and, like everyone else, they were prone to vagaries of fashion.

Progress was made. The First Garden City Company began to pay dividends after a decade; Letchworth continued to grow, more slowly than the promoters hoped, to reach a population of 15 000 – less than half its planned target – in 1938; it was completed only after World War II, ironically aided by government-subsidised decentralisation schemes, and on a slightly smaller scale than originally planned. Ironically, at that point it became the victim of land speculation, from

---

[67] Beevers 1988, 113–114, 131; Fishman 1977, 71–72.
[68] Fishman 1977, 70.
[69] Mumford 1946, 32.
[70] Fishman 1977, 70–71.

Figure 10 *Letchworth. The dream brought to reality, through the Unwin–Parker genius for urban design: neo-vernacular housing, product of close study of old towns, achieves an effortlessly natural look and feel.* Source: Town and Country Planning Association.

which it was rescued by a 1962 Act of Parliament that put its management in the hands of a special corporation.[71] The rise in land values had at last occurred; but it had provoked the response that Shaw had so accurately predicted.

Meanwhile, the Garden City Association had rather markedly changed tack. In 1909, as a direct result of the 1909 Housing and Planning Act, it changed its name and purpose – to the Garden Cities and Town Planning Association, and towards a wider brief: the first aim was now "To promote Town Planning" and "To advise on, draw up schemes for, and promote Garden Cities, Garden Suburbs and Garden Villages".[72] Ralph Neville, in his Presidential address, still put Garden City promotion first.[73] But the Association no longer had the single-minded quality Howard had invested in it.

---

[71]   Miller 1983, 172–174.
[72]   Hardy 1991a, 44–45.
[73]   Hardy 1991a, 45.

Figure 11 *Letchworth. Shopping, the Achilles' heel of the Unwin–Parker partnership: shunted off away from Letchworth's vast central park, it lacks coherence and feels almost like an afterthought – which perhaps it was.* Source: Town and Country Planning Association.

CHAPTER 3

FROM GARDEN CITIES TO NEW TOWNS

At the end of World War I, housing again loomed large on the national political agenda. The surviving servicemen were returning from the horrors of Gallipoli and the Somme, and the new battle-call was Homes Fit for Heroes. One such hero appeared on the cover of a book by Captain Richard Reiss, a member of the Garden Cities and Town Planning Association Council, *The Home I Want*, which featured the memorable message "You cannot expect to get an A.1. population out of C.3. homes". The soldier rejected the grim terraces of by-law housing from which he had clearly come to fight the war, pointing to cottage homes that clearly came from the drawing boards of Unwin, Parker or one of their followers. And Unwin was now a major power in the land: now translated into officialdom, as Chief Architect at the Ministry of Health, he was the driving force behind the hugely influential Tudor Walters report of 1918 which recommended a huge programme of subsidised cottage homes to be built by local authorities. The report, and Reiss's book, made a major impact: Homes for Heroes won Lloyd George victory in the so-called Khaki Election of 1918.[1]

Christopher Addison, a doctor, was put in charge of delivering the programme. There was little doubt that it would be delivered on lines that Unwin had personally advocated in his influential pamphlet of 1912, *Nothing Gained by Overcrowding!*: through cottage homes at densities of which the GCTPA would thoroughly approve, 12 houses to the acre (30/ha). But there was another critical question: where and how?

GARDEN CITIES AND GARDEN SUBURBS

Long before this, Unwin had already changed tack and offended the Garden City purists. When he left Letchworth in 1907, it was to design Hampstead Garden Suburb for Dame Henrietta Barnett. And this effectively split the infant movement ideologically, for though Hampstead Garden Suburb had the appearance of a Garden City and some of its community spirit, it was in every respect a pure commuter suburb; with no industry of its own, it was dependent on a newly opened underground station and was effectively separated from London only by Hampstead Heath.

---

[1] Hall 1988, 66–67.

Hampstead was actually preceded by an earlier garden suburb at Ealing, which Unwin and Parker took over and turned into a delightful mini-Hampstead, and followed by about a dozen others in similar vein, up and down the country, in the years down to 1914. The London County Council architect's department, which was heavily suffused by the Arts and Crafts movement, produced cottage estates in similar vein, culminating in two masterpieces of urban design at Old Oak in west London and Norbury in south London, both completed only after the war. The problem is that all of them were quintessential suburbs, built next to tram or underground lines; they had to be. During the war, Unwin and like-minded architects produced very similar garden suburbs to house munitions workers – in a huge development at Well Hall in south-east London, and on a smaller scale at Kingsbury in north-west London.

But the Garden City purists were at work. Toward the very end of the war, in 1918, a group of them, styling themselves the "New Townsmen" – Howard himself, F.J. Osborn, C.B. Purdom and the Letchworth publisher W.G. Taylor – were arguing that one hundred garden cities should be built under State direction and with the State providing the greater part, 90 per cent, of the £500 million capital required. They specifically denied that this was a Utopian socialist programme, arguing that it was simply a way of achieving industrial efficiency.[2] This was significant: as Howard's biographer has emphasised.

> In sloughing off the embarrassing associations of the past the authors had, in effect, accepted Shaw's early dismissal of the garden city as an alternative to socialism. It was a decisive turning point in the history of the garden city movement. From then on it could be represented as a strand within the town planning movement as a whole, rather than a somewhat esoteric attempt to change the nature of society itself by demonstrating an alternative example within it.[3]

Effectively the argument was Purdom's and the prose was Osborn's; Howard seems to have been content to leave them to it. The pamphlet had another, associated, significance: it was the first published polemical exercise from Osborn, who would become the most effective polemicist in the entire history of the movement (Figure 12).

FJO: THE NEW TOWN POLEMICIST

Frederic James Osborn was born on 26 May 1885 "in a terrace-house in Kennington that my mother's mother let out in lodgings", of Cornish stock.[4] Like Howard, he had a fairly minimal formal education; he said in a radio broadcast in 1967 of his first dame's school, "I reckon I gained nine-tenths of my formal education from Miss Smith, and only about one-tenth from my other two

---

[2] Beevers 1988, 151–154.
[3] Beevers 1988, 154–155.
[4] Whittick 1987, 1.

Figure 12   *Frederic Osborn. "FJO", the crusader and propagandist who persuaded Lord Reith and the Attlee Labour government to embark on the new towns programme at a moment of national bankruptcy.* Source: Town and Country Planning Association.

schools".[5] That early schooling, he said, gave him "the main thing – being able to read early and easily . . . quite by chance I got into a line of creative work – building new towns – that a very miscellaneous assortment of knowledge qualified me for".[6] He might have competed for university, but his parents' income was just

---

[5]   Whittick 1987, 3.
[6]   Whittick 1987, 6.

over the critical limit for a scholarship; so he completed a four-year stretch at his local Council school and left at 15.[7] Then, "a knowing business friend exerted his influence in some 'old pal network' and got me an 'opening' in the City", which proved to be a dead end.[8]

But then, he got a job that would prove relevant, as "clerk–book-keeper to a housing society owning tenement blocks built for workers in the 1870s on a limited-dividend basis – 'Philanthropy and Five per cent'".[9] He said about it, years later:

> That job was like my others in London in that the work left me plenty of time for reading. It also taught me a lot about life in poor parts of London and about housing. It was unsystematic knowledge, but along with many other subjects I picked up in my spare time it helped to make me a plausible candidate for the real jobs I got into later.[10]

Above all, "in business I found that a competent clerk who began as an office boy could eat any average university graduate as a desk-worker or organiser".[11] He joined the Fabians, but the Webbs failed to recognise his emerging talents:

> perhaps they were right to pick out men with a classy air in dress, accent and self-assurance. I was dowdy, spotty-faced, and pathologically diffident – obviously lower middle class and council school . . . by mercy of providence the Webbs didn't take me up. I might have become a probation officer or labour-exchange manager or a borough councillor instead of a new-town builder, and I wouldn't have known what I missed.[12]

Instead, in 1912, at the age of 27, he got the job of secretary and manager of the Howard Cottage Society at Letchworth at a salary of £150 a year; his duties included settling families into new houses and collecting rents. The work expanded as the town grew, and he became involved with estate development generally, getting to know people's likes and dislikes in housing.[13] In a radio broadcast he recalled that his earlier experience gave him a rare gift for "correcting the habit architects and other experts have, of overlooking the desires of the common man and woman".[14] "Architects are mostly in the position of fathers who bolt, leaving the housing manager to hold the baby. So I had to act as a sort of marriage guidance counsellor to architects."[15]

Six years later, he was arguing the case for garden cities – now, interestingly, re-badged as new towns to distinguish them from the *ersatz* variety – to a sceptical minister. Despite Osborn's emerging qualities as a political lobbyist, the Garden

---

[7] Whittick 1987, 7.
[8] Whittick 1987, 7–8.
[9] Whittick 1987, 11.
[10] Whittick 1987, 12.
[11] Whittick 1987, 18.
[12] Whittick 1987, 18.
[13] Whittick 1987, 19.
[14] Whittick 1987, 15.
[15] Whittick 1987, 21.

City movement failed to persuade Addison to adopt its scheme.[16] Addison, the politician, wanted speed; the local authorities could deliver, and so the logic was that the new housing should be built on the edges of the existing cities: the antithesis of the Garden City idea. By 1921, C.B. Purdom in the GCTPA's magazine was complaining about the misrepresentation of the name "garden city" by numerous local councils and speculative builders; "The thing itself is nowhere to be seen at the present date, but in Hertfordshire, at Letchworth and Welwyn Garden City."[17]

HOWARD'S COUP: THE PURCHASE AT WELWYN

Meanwhile, Howard had bypassed them. He had always been dismissive of the other "New Townsmen", Osborn recalled:

> He used to see me off on my missionary tours . . . with comforting words like these: "My dear boy, I hope you have a pleasant trip; but you are wasting your time. If you wait for the authorities to build new towns you will be older than Methuselah before they start. The only way to get anything done is to do it yourself."[18]

He acted on his word; without a word to his colleagues, he had entered into a private correspondence with Lord Salisbury to try to persuade him to sell part of his estate for a second Garden City. Osborn and Purdom both opposed the idea of a second privately sponsored Garden City, and already Purdom was barely on speaking terms with Howard.[19] Then, in May 1919, came the denouement: out of the blue, Howard telephoned Richard Reiss, who had become chairman of the Garden Cities and Town Planning Association, from Hatfield, urgently requesting an immediate meeting: he had just seen an advertisement for the sale on 30 May of a large stretch of land to the north of the Salisbury land. Over a cup of tea at King's Cross, when, as a foreign observer said, he appeared "more agitated than I have ever seen an Englishman", he persuaded Reiss of the need to raise a £500 deposit. In a mad cab ride round the City, Howard persuaded rich friends to bail him out in his astonishing bid for land at Welwyn; when he made the successful bid for a 1500-acre shooting estate at a price of £50 000, he was described as "in an extraordinary state, trembling all over with a fine tremor which was quite beyond his control, and in a cold sweat all over"; Osborn and Purdom were equally shaken.[20] No wonder: as Osborn later put it, "On Friday, 30 May 1919, at an auction sale of part of Lord Desborough's Panshanger Estate, Ebenezer Howard bought 1,458 acres of land for £51,000, without the cash to pay for it."[21]

---

[16] Beevers 1988, 155, 160.
[17] Quoted by Hall 1988, 105–107.
[18] Beevers 1988, 160.
[19] Beevers 1988, 161.
[20] Beevers 1988, 163.
[21] Osborn 1970, 5.

46
—

Somehow the dust settled. A new company was formed and towards the end of October 1920 it decided that it should offer leases of 999 years at ground rents to be revised every 80 years. Thus, Howard's rent–rate principle theoretically survived; however, after a solicitor's advice was taken, and his advice was that this was "not good in law", the company reverted to fixed rents. Howard's central principle had gone into oblivion for a second time. The reason was sheer panic on the part of the Directors, who were liable as individuals for the security of up to £70 000 of bank loans. Instead of the community benefiting, the gains passed to those individuals who bought the housing; all that survived was that the company retained the freehold.[22] As Robert Beevers comments, "The inexorable logic articulated by Shaw in his letter to Neville a quarter of a century earlier finally took hold of his successors at Welwyn and thrust them down the same path despite their commitment to the public good rather than to private profit."[23] Finally, in the depression, encumbered by liabilities of nearly three quarters of a million pounds and on the verge of bankruptcy, the Directors changed the articles so that the Garden City and its inhabitants were excluded from any interest in the company and its profits – the final travesty of Howard's ideas, after his death.[24] After World War II the town was effectively nationalised, becoming one of the first wave of London new towns.

So the vital financial principle was abandoned. And Welwyn was compromised even more basically, because it was far from a self-contained cooperative commonwealth; only 20 miles (32 km) from London, 15 miles (24 km) nearer than Letchworth, from the start half the population were prosperous commuters to London, only 35 minutes away when the station opened in 1925. To this day, ironically, the offices of the Town and Country Planning Association are hung with 1920s posters advertising the advantages of commuting from the Garden City. The remainder of the population worked for the Welwyn Garden City Company's various subsidiaries and lived in council housing, so in effect this was a company town on the model of Bournville or Port Sunlight; not what Howard had in mind at all, though by now he was the second Garden City's most distinguished resident.[25]

He died in 1928. Not long before, he had said

> When we have done our work, it is not well to linger on. I do *want* to live a bit longer, because I have something definite I want to do; but if I am taken sooner than I want to be, it will still be for the best, not only for me, but for the work.[26]

That something, of course, was Welwyn, bought a mere nine years earlier when Howard had reached the age of 69; a wonderful old man's whim. The directors of the company may have ruminated more than once on those words; but in 1998

[22] Beevers 1988, 169–170.
[23] Beevers 1988, 175.
[24] Beevers 1988, 175.
[25] Beevers 1988, 174.
[26] Macfadyen 1933, 157.

Welwyn is a magnificent memorial, and as visitors come from the station through the Howard Shopping Centre they doubtless reflect on it (Figures 13 and 14).

THE LONG WAIT

During virtually the whole of the 1920s and 1930s, it seemed, the Garden City–New Town movement was becalmed.

Osborn was not quite as old as Methuselah – as Howard had so memorably predicted – when the British government finally acted to build new towns: in 1946, when the New Towns Act was passed, he was 61, almost as old as Howard when he had made that statement. In fact, he had nearly 30 more great campaigning years before him. But between the wars, many faithfuls must have reflected on those words. Campaigning for new towns in those years, as one can sense from the yellowing pages of the magazine *Town and Country Planning* in the 1930s, was very much crying in the wilderness.

Garden Cities alone would not serve as a rallying-cry. Particularly after Osborn's appointment as Secretary to the GCTPA in 1936, the campaign broadened, culminating in the change of name to the Town and Country Planning Association, approved by members in 1941. During the late 1930s, Osborn skilfully developed a coalition with preservationist forces, and even affiliated with the Council for the Preservation of Rural England; Professor Patrick Abercrombie (Figure 15) was active in both bodies. Following this up, in 1939 the GCPTA called for a "Planning Front" or lobby of these different bodies; and the next year the 1940 Council, chaired by Conservative landowner Lord Burleigh, essentially did this. It campaigned for regional planning to counter suburban sprawl; but to little or no effect, as confessed in an editorial of December 1939. As a despairing Raymond Unwin had reminded an audience in the early 1930s, the entire population of Letchworth and Welwyn was about 24 000, equivalent to 12 weeks' growth of London. Responding to the opportunity created by the creation of the Barlow Commission on the Distribution of the Industrial Population, in 1938 the GCPTA submitted a superbly researched and well written 43-page memorandum (from Osborn, of course), which – backed up by extremely astute lobbying on Osborn's part – undoubtedly helped powerfully to shape the Commission's conclusions.[27] The GCPTA/TCPA also developed a very strong campaigning line against high-rise, high-density solutions, then being advocated by modernist architects; and it resolutely criticised and pressured Abercrombie over what Osborn felt to be a half-hearted attitude to decentralisation in the 1943 County of London Plan, which bore fruit in the much more radical proposals in the Greater London Plan the year after.

There were enthusiasts in both the main British political parties: on the Conservative side Neville Chamberlain, Minister of Health in the mid-1920s, was one, and would undoubtedly have supported the programme if economic

---

[27] Hardy 1991a, 167, 174, 177, 186, 195, 199, 201, 214–215, 255–256, 259, 269.

Figure 14 *Welwyn Garden City. Louis de Soissons' formalistic handling of the shopping centre via a carfax, leading off his central processional way, to the station; now terminated by the Howard Centre, the new shopping centre.* Source: Peter Hall photograph.

conditions had been better. On the Labour side, in a special feature in the Association's journal in 1937, when notable celebrities pledged themselves to the cause, one statement came from the recently elected leader of the Labour Party, one Clement Attlee. When he became Prime Minister, as he was wont to do, Attlee delivered on his promise. Even in 1945 when the Labour Party came to power, there was far from unanimous enthusiasm, and it took all of Osborn's powers of persuasion to bring the legislation to reality. The new Minister for Planning, Lewis Silkin, a sophisticated lawyer, had been a member of the London County Council and was not originally enthusiastic; he had to be won over. A government faced with unprecedented demands on resources, as the Attlee government was at war's end, was bound to be frightened by such a commitment.

But the claims of postwar reconstruction, including the provision of "homes fit for heroes" – a slogan at the end of World War I, on which the government of the day had reneged – proved compelling: this time, the government must get it right. In particular, Sir Patrick Abercrombie's famous 1944 plan for Greater London, commissioned by the wartime government, had identified and quantified an enormous problem of overspill: if the bombed and blighted areas of east and

Figure 13 *Welwyn Garden City. Post-World War I, Louis de Soissons' translation of the garden city genre into neo-Georgian; but the design is still organic, and feels effortlessly right.* Source: Peter Hall photograph.

Figure 15 *Patrick Abercrombie. British planning's first great academic–practitioner, who managed to reconcile the aims of the TCPA and the CPRE; his 1944 Greater London Plan brilliantly traded off urban containment, countryside protection and the creation of new communities.* Source: Town and Country Planning Association.

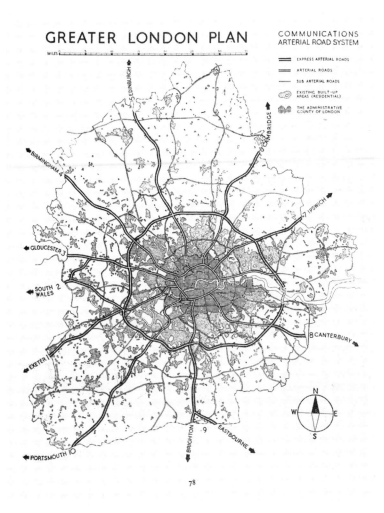

78

Figure 16 *The Greater London Plan 1944. Inner London is thinned and reconstructed, London's growth is stopped, new towns are built to house the overspill, and country towns are expanded, all against a background of continuous open countryside. Source: Abercrombie 1945.*

south-east London were to be rebuilt at adequate standards, with small gardens for families with children, then over one million people would have to move out of London in planned overspill schemes. And, since the Abercrombie plan called for London's further expansion to be limited by a 10-mile (16-km)-wide green belt, there was really no alternative to the solution he proposed: a combination of completely new towns and planned town expansions, outside the green belt and thus at a minimum distance of 20 miles (32 km) from the centre of London (Figure 16).

Osborn already began to work on preparing the political ground at this time; as early as January 1944 he had been instrumental in persuading the chief technical adviser of the Ministry of Town and Country Planning to set up an internal committee on new towns, and submitted a long memorandum to it; in early 1945 he then persuaded Silkin to co-operate in a joint TCPA–LCC study of the problem. Thus, when Silkin became Minister after the election, he was half-persuaded and appointed an expert committee, chaired by the redoubtable Sir John Reith – ex-Director General of the BBC and himself briefly a Planning Minister in the wartime coalition government – to consider how to build new towns. Basing their work on the solid foundation left by the Pepler Committee, they recommended that the new towns should be built not by local authorities but by public development corporations directly financed by Treasury grant; they would require a huge front-end investment, but would later yield big commercial returns to the government.[28]

Michael Hebbert shows that the committee rejected any private sector role out of hand: "In the reconstruction climate of the mid-1940s it was feasible to contemplate a more radical ring-fence against market forces than in the pioneer garden cities."[29] The Committee had only two private sector representatives, Sir Malcolm Stewart (chairman, London Brick Company) and L.J. Cadbury (chairman, Cadbury Brothers Ltd). Belatedly it agreed that it should consult private developers, who were enthusiastic: Taylor Woodrow told them they could build a new town for 50–70 000 in five years, with two-thirds of houses sold to inhabitants on 99-year leases; financial institutions affirmed that investment could be available. After some reluctance, the Committee's interim report did allow for the possibility of Authorised Associations, operating on a limited-dividend basis under the 1932 Town and Country Planning Act. But the minister, Lewis Silkin, announced in March 1946 that he would reject this, and the New Towns Bill provided only for public development corporations.[30]

Then the Reith committee identified the stark political reality: to get the new towns started quickly, in order to make a start on housing even part of the overspill, construction would have to be taken out of local authority hands. The counties and municipalities outside London were unlikely to have much enthusiasm for the project, and could even be downright hostile. Ironically, 35 years later, the same arguments persuaded the first Thatcher government that only development corporations would effectively do the job of regenerating Britain's inner cities.

THE POSTWAR NEW TOWNS

The New Towns Act for England received Royal Assent, passing it into law, on 1 August 1946; the first new town, Stevenage, was designated on 11 November 1946.

---

[28] Hardy 1991a, 279–280.
[29] Hebbert 1992, 169.
[30] Hebbert 1992, 170–173.

For Osborn and the TCPA, it was an immense triumph: "The cause of garden cities, albeit now with a new name, had been advanced from its origins within the covers of a cheap book with a readership of late-Victorian 'cranks', to the status of an Act of Parliament with the prospect of a programme for the immediate implementation of new towns in various parts of the country."[31] Osborn himself reflected on his personal influence:

> I think . . . that I personally have been a decisive factor in the evolution of the new towns policy and that this evolution is extremely important historically. I mean no less than without my fanatical conviction and persistent work in writing, lecturing and especially lobbying, the New Towns Act of 1946 would not have come about, at any rate in that period.[32]

In the half-century after 1946, Britain built 28 new towns, with a combined population of 2 254 300 people at the 1991 Census (or 3 per cent of the entire national population) as against only 945 900 at the start. Thus, 1 308 400 people were housed in planned new communities; all with accompanying employment – some 1 110 000 jobs, against 453 000 at the start, a net increase of 646 600 (Figure 17).

These new towns were built in two concentrated bursts of activity. The first came from 1946 to 1950, immediately after passage of the Act, under the Attlee Labour government, and it produced exactly half of the 28 new towns: eight in a ring around London, all between 21 and 35 miles (34 and 56 km) from the centre; one in the English Midlands, in the steel town of Corby; two in North East England, Newton Aycliffe and Peterlee; one in South Wales, in Cwmbran; and two in central Scotland, at East Kilbride and Glenrothes. Then there was a long break during the 1950s, when successive Conservative governments – led by Winston Churchill, Anthony Eden and Harold Macmillan – set their face against further new town designations; only one, Cumbernauld in Scotland, started during this time. And then there was another big burst, from 1961 to 1970, when – first under Macmillan's Conservative administration, then under Harold Wilson's Labour government – 13 more new towns were started: three more for London, but this time much more distant, between 60 and 80 miles (96 and 130 km) away; two for the Midlands, Redditch and Telford; four for the North West, Runcorn, Warrington, Skelmersdale and Central Lancashire; another for the North East, Washington; one more for Wales, Newtown; and two more for Scotland, Livingston and Irvine. The last designation, that for Central Lancashire, came in 1970.

The record also shows that the new towns differed not only in their geography, but also in size and function. In 1991, they ranged in size from tiny Newtown in Mid-Wales, with only 11 000 people, all the way up to Central Lancashire with 255 000. But that last was a statistical fluke, because Central Lancashire's growth was cut artificially short just after designation and the net growth was a mere

---

[31] Hardy 1991a, 282.
[32] Quoted in Hardy 1991a, 283.

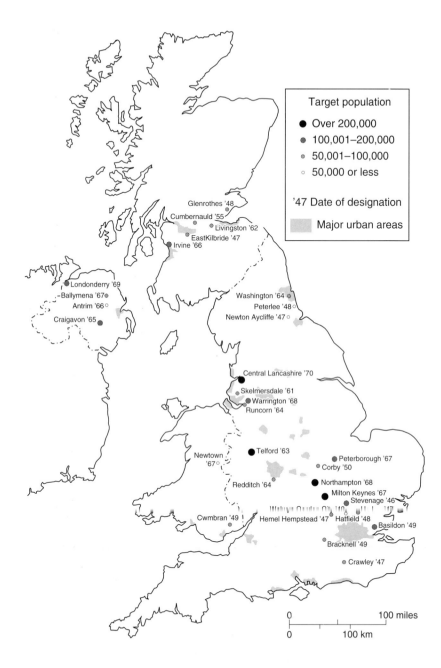

Figure 17 *The British New Towns. The permanent legacy of FJO's campaigning: nearly 30 new communities, housing a million people, built on sustainable principles with jobs and services within easy reach.* Source: Hall 1992.

21 000 people; more spectacular was the growth of Stevenage from 6700 to 75 000, or Milton Keynes from 40 000 to 143 000, or Basildon from 25 000 to 158 000.

Functionally, all but three of the British new towns fall into well-defined groups: 11, over one-third of the total, were designed to receive London overspill; another 10 were started to receive overspill from the great conurbations of the West Midlands, Merseyside, Greater Manchester, Tyne and Wear and Greater Glasgow; and a small group of four – two in England, one in Wales, one in Scotland – were designed to aid regional regeneration of coalfield or heavy industrial areas. The three odd cases are Corby in the East Midlands, built to provide housing and services for a nationalised steel plant; Glenrothes in Scotland, similarly for a new coalfield; and Newtown in Mid-Wales, built to assist the regeneration of a declining rural area.

An impressive programme, then, but one element of Howard's vision was forgotten. Bigger, later on much bigger, than Howard's Garden Cities, the postwar new towns are essentially freestanding; they do not form Social City clusters. There is however one major exception, and ironically – whether by accident or design, and one can suspect the latter – it is in mid-Hertfordshire, where Howard had begun both his experiments in Garden City building. To Letchworth, 35 miles (56 km) from London, and Welwyn, 20 miles (32 km), the postwar planners added the first designated new town at Stevenage, a site that had appeared repeatedly on the Garden City enthusiasts' maps of prospective sites since the 1920s, neatly located about halfway (28 miles, 45 km, from London). Then they added another around the de Havilland factory at Hatfield, a mere 18 miles, 29 km, away; they coupled this latter with a state takeover of Welwyn Garden City and its most distinguished resident, F.J. Osborn, and brought the city under the same development corporation as Hatfield. Thus they achieved Howard's final dream: Mid-Hertfordshire became a 17-mile (27-km) linear realisation of Howard's principle of the Social City, the group of slumless smokeless cities, self-contained but linked by rail rapid transit, that he had illustrated so graphically in the first edition of *Tomorrow!*; the only such realisation in the world. In the 1990s, with associated settlements like Hitchin, it has a population of about 250 000: almost exactly the same as in the famous diagram. Some of the constituent units are much bigger than Howard advocated: Stevenage has a population of about 75 000, because planners decided after the war that a town of that size was necessary to provide an adequate range of jobs and shops and services, and if one compares Stevenage with Letchworth they are surely proved right. The increase in size means, of course, that Stevenage is not a walk-to-work town, though by way of compensation it provides very well for bikes. In this realisation Howard's Inter-Municipal Railway, a light rail system, was replaced by some very heavy existing rail in the form of the London and North Eastern Railway, which in 1947 became part of British Railways.

Building the new towns evoked opposition, sometimes passionate, just as later the activities of the London Docklands Development Corporation did. It was a neat mirror image, for in the late 1940s the shire counties and districts around London were invariably Conservative, and the idea of importing thousands of working-class Londoners – who would invariably be Labour voters – was anathema. At Stevenage, the enraged locals tore down the train station nameplates and replaced them with ones saying SILKINGRAD (see Figure 53); when Silkin came to a public meeting there was uproar, and when he left the meeting he found the tyres of his car had been let down.[33]

Slowly, however, the new towns got started – though often the start was symbolic, for the economic crisis of 1947–48 meant that all social programmes had to be cut back. Ironically, the bulk of the construction of the so-called Mark One new towns took place under Conservative governments whose announced plan was to terminate the programme. They started only one new town in 10 years: Cumbernauld, to meet Glasgow's pressing overspill problems. In England they preferred to proceed under the Town Development Act of 1952 – a measure long foreseen, and actually prepared under the previous Labour government – which would provide the other arm of the Abercrombie programme: planned expansions of existing small country towns, through agreement between exporting and importing authorities, with Treasury funding for basic infrastructure. But that programme started very slowly, and it proved much harder than expected to forge the necessary agreements; so the new towns programme had to proceed in its place. Eventually, the expanded towns programme also yielded a big housing programme, and in southern England some of the larger planned expansions of the 1960s – Basingstoke, Andover and Swindon west of London, and Wellingborough north of the capital – are almost indistinguishable in scale or style from the pure new towns.

The opposition of local authorities, and the lack of enthusiasm of Conservative governments, did however affect the geography of the new towns programme. An oddity was that virtually all the Mark One overspill new towns were built around only two cities, London and Glasgow. It was true that together they did present housing problems on an unusually large scale: London because it was so big (and also because it had been bombed), Glasgow because it had the worst housing conditions of any British city. But Birmingham, Manchester and Liverpool also had huge rehousing deficits after the slum clearance programme recommenced in earnest in 1955, and at this stage no new towns were forthcoming for them. A relevant fact was that there was an atmosphere of suspicion between the cities and their neighbouring shire counties. Manchester tried to develop a new town at Lymm in neighbouring Cheshire, Birmingham tried to develop a satellite community at Wythall in neighbouring Worcestershire, but after extended public inquiries in the late 1950s both lost the battle.

The Mark One new towns started between 1946 and 1950 are very much cut from the same cloth and the same pattern book. For there was an orthodoxy, which had developed out of the original Howard formulation and had been honed through various refinements on both sides of the Atlantic. This occurred most notably through the concept of the neighbourhood unit, developed by Clarence Perry in the New York Regional Plan of 1929–31, and its application by Barry Parker at Wythenshawe in Manchester, sometimes called the third garden city, in the early 1930s. A planned town centre, which in Stevenage was pedestrianised from the beginning and in the other new towns somewhat later, was connected by

---

[33] Collings 1987, 15, 19.

landscaped roads with the neighbourhood units, which were designed around schools and district centres. Industry was carefully segregated from residential areas, and was located close to planned motorway or trunk road interchanges. There was generous park space, and often provision for walking or cycling away from the traffic routes (Figures 18 and 19).

But, after 1955, a new and unexpected element entered the scene of policy formulation: just as the government started again on slum clearance in the cities and called on shire counties to establish firm green belts around the major cities, the birth rate began to rise. The demographers had not predicted it, and by the late 1950s it brought a crisis in urban policy: the major provincial cities were running out of land for redevelopment, the expanded towns programme was going too slowly to absorb the overspill, and there was a real risk that their slum clearance programmes would grind to a halt. It was this combination of forces that caused a major reappraisal around 1960–61, and with it the resumption of the new towns programme: Skelmersdale outside Liverpool, first of the English Mark Two new towns, was announced in 1961, followed by Dawley outside Birmingham (designated 1962 and later enlarged and renamed Telford in 1963).

These Mark Two new towns were for the most part designated to remedy a glaring gap in the previous programme: the lack of provision for overspill from the great provincial cities and their conurbations. Thus Dawley–Telford and Redditch were started for the West Midlands conurbation around Birmingham; Skelmersdale and Runcorn for the Merseyside conurbation around Liverpool; Warrington and Central Lancashire for Greater Manchester; Washington for the Tyne and Wear conurbation around Newcastle upon Tyne; and two further new towns, Irvine and Stonehouse, for Greater Glasgow. Generally, with the exception of Irvine, these tended to be relatively close to their parent conurbations, typically only 12 miles (19 km) distant – a measure of the smaller size and compact shapes of these provincial agglomerations.

The new towns of this era, usually dubbed the Mark Twos, were quite often conceptually different from their Mark One equivalents. They tended to follow the pioneer, Cumbernauld in Scotland, in being planned much more consciously for the private car. Because this was already beginning to change shopping patterns, there was now a greater emphasis on the town centre, which might become an enclosed mall. To try to handle the pedestrian/car conflict, planners adopted different devices: in Cumbernauld, and in the never-built London County Council new town at Hook in Hampshire, they connected residential areas directly to the centre via pedestrian pathways, diverting the cars on circuitous motorway-scale highways; at Runcorn, they provided a bus service running on a segregated busway, still a novelty in the 1990s (Figures 20 and 21). Generally, though not universally, their adherence to the neighbourhood unit began to show signs of weakening.

Then, in the late 1960s, came three further new towns for London – often styled the Mark Three new towns. These were super-new towns, designed for much larger populations than the Mark One versions: typically in the range 170–250 000. All were much further from London: Milton Keynes was 60 miles (96 km) away, Northampton 70 miles (113 km) distant, and Peterborough 80 miles

(130 km) from London. This, it was argued, was necessary to guarantee their self-containment in an age of rapidly rising mobility. But all were on the main transport corridors linking London with the Midlands and North, and so had excellent connections both with London and with the other great cities. And while one, Milton Keynes, was essentially a greenfield new city incorporating some small pre-existing towns and villages, the other two – Northampton and Peterborough – were enlargements of medium-sized ancient county towns, with the idea of providing a built-in level of services from the beginning. All continued the Mark Two tradition of providing for a very high level of car mobility, through generous highway provision. Because of their size, they could not reject the neighbourhood principle, and though the Milton Keynes planners tried to replace it by a concept of spatial ambivalence, in which people would be free to patronise one centre or another more or less indifferently, nevertheless these centres were just as much a definite part of the structure as was the giant central shopping mall (Figures 22 to 24).

These Mark Three new towns for London reflect a very conscious attempt by the national planners of the 1960s to adapt Howard's principles to the new realities of postwar Britain. Both Howard, and the postwar new town planners, aimed to build their new settlements deliberately outside London's commuter range. They must have already known that it was an impossible dream, because Welwyn developed as a commuter town from the 1920s; even F.J. Osborn, famous new town advocate and Secretary of the Town and Country Planning Association, was one of them. Just at the time of the designation of the Mark Two towns, Ray Thomas made a detailed analysis of the performance of the Mark One generation on the basis of the 1966 Census. He was able to show that they remained much more self-contained than equivalent older towns at similar distances from London.[34] But when Michael Breheny came to rework the figures 20 years later, he found that they were losing this characteristic: the London commuter belt had expanded into the territory they occupied, and besides there was a widespread growth of commuting as car ownership grew.[35] It was partly because planners anticipated this fact that, in the 1960s, the planners put the Mark Two new towns much further from London and made them even bigger, so that they could offer all the jobs and services that you would associate with a major provincial city like Leicester or Plymouth. But, of course, this meant further sacrifice of the walk-to-work principle, which rather got lost in those years when even planners worshipped at the altar of the automotive god.

The designation of the Mark Two and Mark Three new towns reached a peak under the Wilson government around 1967–68, and their construction took place

---

[34] Thomas 1969.
[35] Breheny 1990.

Figure 18 Harlow. Typical "Mark One" new town planning on the neighbourhood principle, with houses grouped around local shops and schools. Source: Town and Country Planning Association.

Figure 20 *Cumbernauld. The only new town started for a decade, its "Mark Two" design for the motor age broke sharply with the neighbourhood principle, clustering houses along pedestrian paths with direct access to a single shopping centre.* Source: Town and Country Planning Association.

Figure 19 *Basildon. Another of the "Mark One" new towns for London, which provided ideal living environments for the postwar "baby boom".* Source: Town and Country Planning Association.

mainly in the 1970s and early 1980s. They still remained in the mould the Reith Committee had set for them in 1946, dominated by rented public housing: even around 1970, only one in ten completions was privately built. But then a rapid change occurred, and by the end of the 1970s half of completions were private, rising to 90–95 per cent from the mid-1980s.[36]

However, already, in one of those strange cycles of policy that seems to characterise Britain, enthusiasm for the programme was weakening. A major reason was the British government's slow acceptance that there was indeed an inner-city problem, and that its causes were deep and structural: the cities were losing both people and jobs, and this process was continuing long after the policies said it should cease. And underlying that, in turn, was a long process of deindustrialisation and associated contraction of all kinds of goods-handling employment in older industrial areas and older dockland areas in the major cities.

A major turning-point in policy – some would argue the major turning-point in postwar British planning history – came in 1977 with the publication of consultants' reports on the problems of inner London, Birmingham and Glasgow, followed by a government White Paper and by the Inner Urban Areas Act of 1978. Essentially, this transferred resources from the new and expanded towns programmes to urban regeneration. It had already been trailed in Scotland, where in 1974 Stonehouse new town was abandoned and the funds effectively switched into a major regeneration programme, the Glasgow Eastern Area Renewal or GEAR. Now, under the Callaghan Labour government, the emphasis was on partnership programmes to assist the inner cities. After 1979 the Thatcher government would change the policy instruments, introducing Enterprise Zones and Urban Development Corporations, but would essentially retain the new emphasis: this was a major governmental U-turn.

There was a strange and partial exception: in the mid-1980s, the idea of so-called new communities, essentially new towns initiated and constructed by private enterprise. Nine, later ten, of the biggest volume builders joined together in 1983 as Consortium Developments to pursue this policy, apparently believing themselves encouraged by government. Very few of their and others' schemes, some 30 in number, came to fruition; some because they were poorly sited (e.g. Tillingham Hall in the London green belt); some because of bitter local opposition. The strangest case was that of Foxley Wood some 45 miles (72 km) west of London: one Secretary of State said (in the strange legal language used on such occasions) that he was "minded to approve it"; his successor turned it down. The fact was that, despite the ideological enthusiasm of Thatcher governments for such ventures, they proved politically unpopular with local voters: NIMBY, Not In My

---

[36] Hebbert 1992, 175.

Figure 21 Runcorn. A "Mark Two" design from the mid-1960s, with segregated busways converging on an enclosed shopping mall, Shopping City; hit by economic vicissitudes in the 1990s, it remains an amazingly innovative design. Source: Town and Country Planning Association.

Figure 23 *Milton Keynes. Early housing, regimented, geometrical and unpopular: ironically, it represented a major departure from the original Llewelyn-Davies plan.* Source: Town and Country Planning Association.

Backyard, became their slogan and it proved only too effective. In any case, new town purists were not slow to point out that they were not truly new towns at all, but long-distance garden suburbs, the 1980s equivalent of Hampstead Garden Suburb in London: they were for the most part true dormitory communities, lacking substantial local sources of employment.

The central problem of the privately built new communities, as Michael Hebbert has pointed out, is that the special costs of new urban development are financed out of enhanced land values. Developers in the early 1980s seem to have expected that they would be granted planning permission on protected farmland of low current value: the "economics depended, in effect, on bucking the planning system".[37] But, Hebbert pointed out, that game could only be played once or twice: as soon as the precedent was set, land prices would rise and profits shrink. And the government reacted by stating that private developers could obtain planning permission only for developments envisaged in local authority plans – that is,

---

[37] Hebbert 1992, 182.

Figure 22 *Milton Keynes. Most celebrated of all the new towns, with the archetypal "Mark Three" design of the late 1960s: huge, and uncompromisingly car-oriented.* Source: Town and Country Planning Association.

Figure 24 *Milton Keynes. Willen, district centre for one of the last completed neighbourhoods; a triumphant re-affirmation of the principles of the original plan, and an illustration of what the entire city could and should have been.* Source: Peter Hall photograph.

areas where the owners had already made windfall gains and thus lowered potential development profit.[38]

Hebbert shows that this problem is not new:

> When Ralph Neville, chairman of the First Garden City Company Ltd., was interviewed by a parliamentary committee in 1904, he took a more sober view of the economics of the enterprise than did Tom Baron in an adjacent committee room sixty years later: "You could not go on indefinitely in this matter by private enterprise, it would be impossible; first of all, because you could not get the sites at a reasonable price again. We have had enormous difficulties ourselves . . . but the thing started, the idea taking on, the price could be prohibitive."[39]

Neville hoped for a public–private partnership – as was envisaged by Frank Taylor before the Reith Committee in 1945, and as actually used at Cramlington in Northumberland and South Woodham Ferrers in Essex – and this, Hebbert concludes, seems the most promising avenue for building future new towns.[40]

THE LESSONS OF THE NEW TOWNS

So, in the 1990s, the new towns appear truly as a piece of British history, on which the last page has almost turned. But perhaps the old book is about to be reopened, and a new chapter is about to be written.

If it is, or when it is, two key features of the new towns programme should be remembered. The first is that as a programme for deconcentrating people and jobs from London and the other major conurbations, the new towns were outstandingly successful. True, as seen, in quantitative terms they took only a minuscule share of the total growth of the British population over the half-century during which they were being built. But they showed the way. Coupled with tough negative planning controls that established and then maintained green belts around the major metropolitan areas, and a broad green backcloth beyond those, they established what became a distinctive feature of postwar British planning: the system of towns against a backcloth of open country, as Raymond Unwin described it when it was still a dream. In a sense, old towns that grew – (some of them by planned overspill under the 1952 Act like Basingstoke, Andover, Wellingborough and Haverhill, many more by the normal processes of growth and the normal operations of the planning system) – mimicked the new towns: throughout Britain, and nowhere more than in the pressured South East, the pattern is one of small- or medium-sized towns, against a background of continuous open country. This perhaps is the greatest achievement of the system, and the model of the new towns has a lot to do with it.

The other main point, which is too often overlooked, is that at the start the New Towns were a conspicuous financial success, even though later on they suffered

---

[38] Hebbert 1992, 182–183.
[39] Quoted in Hebbert 1992, 183.
[40] Hebbert 1992, 183.

huge artificial burdens placed on them by the Treasury. Here the work of Ray Thomas of the Open University is uniquely valuable.[41] He shows first that the development corporations were able to buy land for housing at amazingly low prices: half of one per cent of the average cost of a house in the 1950s, a little over one per cent in the early 1970s, and secondly that they built up huge surpluses – to the extent that by the mid-1970s Harlow Development Corporation was acting as a kind of merchant bank, lending to the local water authority. The Treasury's response was to appropriate Harlow's surplus funds, meaning that later in the 1970s the new town had to pay one hundred times as much for land as it had a quarter-century earlier – an accurate measure of the way the town had created its own land values. When the development corporation handed over its commercial assets to the Commission for the New Towns, in 1980, they were yielding 12 per cent per annum in rents.[42]

But, Thomas shows, such wonderful results were achieved only by the first-generation new towns of the 1940s. Later new towns such as Milton Keynes never made a profit; in all, they accumulated a deficit of some £1500 million. There were two main reasons: they lost control of housing rents under the 1972 Housing Finance Act, and even more importantly in the 1970s they faced ruinous interest rates: 18 per cent over a 60-year term. In the late 1970s Milton Keynes was having to pay an average of 13 per cent, but was getting a return on its industrial and commercial property of only about 8 per cent – in stark contrast to Harlow, which was paying only 5 per cent and getting 12 per cent or more. And on top of that, a town like Milton Keynes saw most of its development take place in the relatively depressed property markets of the late 1970s and early 1980s.[43]

All that might suggest that the later new towns were a drain on the public purse. Nothing could be more wrong; their assets appreciated massively, both through the creation of new urban values, and through inflation. During the years 1986 to 1992, when Milton Keynes Development Corporation was revaluing and selling its assets before selling the rest to the Commission for the New Towns, its accounts indicate large profits. It also handed over land which it had bought for under £25 million or just over £2000 per hectare, but whose market value in 1986 would have been one hundred times that amount; the Commission for the New Towns still owned 2000 hectares of that land in 1994.[44]

Ray Thomas concludes that the new towns programme suffered grievously from a sharpening focus on short-term economic management on the part of the Treasury. But private fundholders do not behave like that at all. Thomas concludes:

> A major demographic feature of Britain and other advanced industrial countries is an ageing population and the growth of pension funding. A significant part of the funds available for investment is looking for long-term capital appreciation. A revived new

[41] Thomas 1996.
[42] Thomas 1996, 306.
[43] Thomas 1996, 307.
[44] Thomas 1996, 308.

towns programme could well look in this direction, rather than to the Treasury, for financial support. The implication is that new town development corporations would have to find ways of combining commitment from government and funds from the private sector.[45]

That was also the conclusion of Michael Hebbert's historical review. It will need pondering if, or when, any government contemplates starting a new towns programme all over again.

---

[45] Thomas 1996, 308.

CHAPTER 4

PLOTLANDS: THE UNAUTHORISED VERSION

In the early decades of the twentieth century, electric tramways, the "gondolas of the people" as Richard Hoggart called them, had established themselves as *the* mode of popular transport in British towns and cities, cheap and reliable even if noisy and uncomfortable. They had the advantage of motive power at cost price since the municipal tramway operator was usually also the power generator. In 1920 the reverse side of London County Council tram tickets carried an advertisement for plots of land at "The Garden City by the Sea".

This was the slogan applied by its speculative developer to Peacehaven, the estate on the south coast between Newhaven and Rottingdean, where Londoners were invited to build the bungalow of their dreams on very cheap freehold plots. This development had nothing in common with the Garden City concept but was gleefully ridiculed by Thomas Sharp, the most prolific planning propagandist of the 1930s:

> The "reductio ad absurdum" of the garden-city is its extension to absurdity, and of this, unfortunately, innumerable examples exist. The worst in England is Peacehaven, which has rightly become a national laughing-stock . . . It is indeed a disgusting blot on the landscape.[1]

Peacehaven was simply the most notorious of those settlements for which town planners have invented the useful word 'plotlands'. It is a shorthand description for those areas where, in the first 40 years of this century, farmland was divided into small plots and sold, often in unorthodox ways, to Londoners wanting to build their holiday home, country retreat or would-be smallholding. The word evokes a landscape consisting of a gridiron of grassy tracks, sparsely filled with bungalows constructed from army huts, railway coaches, shanties, sheds, shacks and chalets, which when left to evolve on its own, slowly becomes like any other ordinary suburban landscape, leaving only a few clues to its anarchic origins.

By 1939 this plotland landscape was to be found in pockets across the North Downs, along the Hampshire Plain, and in the Thames Valley at such riverside sites as Penton Hook, Marlow Bottom and Purley Park. It was interspersed among the established holiday towns on the East and West Sussex coast at places like

---

[1] Sharp 1932, 143.

Shoreham Beach, Pett Level and Camber Sands and, most notoriously, at Peacehaven. It crept up the East Coast, from Sheppey in Kent to Lincolnshire, by way of Canvey Island and Jaywick Sands, and it clustered inland all across South Essex (Figure 25).

The plotland phenomenon was not confined to the south-east of England. Every industrial conurbation in Britain once had these escape routes to the country, the river or the sea. For the West Midlands there was the Severn Valley, the Wye Valley and North Wales, for Liverpool and Manchester the North Welsh coast and the Wirral, for Glasgow the Ayrshire coast and even the bonny banks of Loch Lomond. Serving the industrial populations of the West Riding towns and cities were sites along the Yorkshire coast and the Humber estuary, and for those of Tyneside and Teesside there were the nearby coasts of Northumberland and Durham.

It was as though a proportion of the population was obeying a law of nature in seeking out a place where they could build for themselves. But it is certainly worth remembering that, when the plotland phenomenon began, most families in British cities and towns were only one or two generations away from rural life.

A series of factors made the plotlands. The first was the same economic fact that influenced Ebenezer Howard, the price of rural land. There is an old saying that land loses its value long before it loses its price. But agricultural decline as a result of cheap imports, which had begun in the 1870s and continued (with a break during World War I because of submarine blockades) until 1939, encouraged the buying and selling of bankrupt farms at throwaway prices. In 1913 you could buy land in Kent at £10 an acre (£4/ha) with a £1 deposit, or a plot on Canvey Island in Essex for 11s.6d. (57.5p). The break-up of landed estates after the Liberal government's doubling of death duties, together with the slaughter-rate among sons and heirs of landowners in World War I, added to the pressure among sellers to seek a multitude of small buyers in the absence of a few large ones.

A second factor was the spread down the social scale of the holiday habit and the idea of the "week-end". The Holidays With Pay Act of 1938 affected 18.5 million employed workers (and consequently their dependents), nearly 11 million of whom were to receive holiday pay for the first time. Those who previously took a holiday paid for by savings, were likely to seek a cheap one, and a glance at *Dalton's Weekly* in the 1930s would show that, apart from a tent in a field, the cheapest holiday advertised was to rent someone else's plotland bungalow.

Another factor was the accessibility of cheap transport. There was the incredible ramification of railway branch lines, reaching places for which the station had become an unexpected link with the outside world, but also the pleasure boat industry, fighting back against railway competition in the holiday trade, not merely up the Thames, but also along the Essex and Kent coasts. A final, decisive factor was the gradual democratisation of private motoring.

The fourth factor is best summed up as the growth of the cult of the great outdoors. This had several aspects. One was belief in the health-giving qualities of fresh air, as a defence against such scourges of urban life as bronchitis and tuberculosis. Another aspect was the pursuit of popular riverside and seaside sports such as fishing, boating and sailing. Yet another aspect was the attraction to

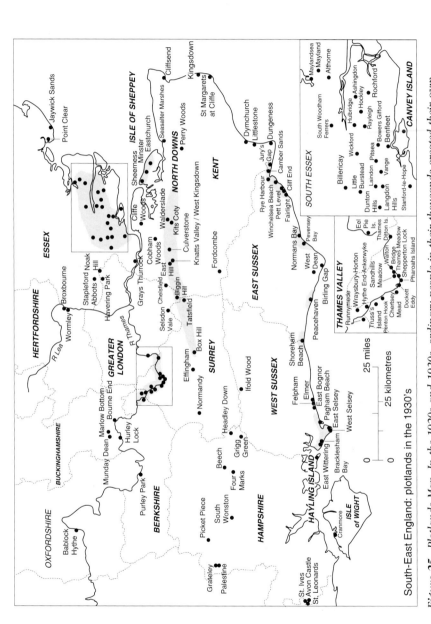

Figure 25 *Plotlands Map. In the 1920s and 1930s, ordinary people in their thousands created their own communities all over South East England; today, they are either obliterated or transformed beyond recognition.* Source: Hardy and Ward 1984.

South-East England: plotlands in the 1930's

dwellers of "the simple life", whether in a country cottage with three acres and a cow, or as a long-distance commuter.

There was, finally, the idea of a property-owning democracy. At the end of the twentieth century, the major mode of tenure in Britain is the owner-occupied house. When the century began, 90 per cent of households, whether rich or poor, rented their dwellings,[2] and throughout the twentieth century the attraction of possessing a few square yards of England has had its appeal. Long before a minor Conservative politician coined the phrase about property-owning, one plotland entrepreneur, Frederick Francis Ramuz, twice mayor of Southend-on-Sea, who operated as The Land Company, was advertising in 1906 that "Land Nationalisation is Coming", meaning that the dominance of the absentee landlord would be replaced by every family owning its portion of our common birthright. Like the developer of Peacehaven, Mr Ramuz claimed that on his sites at Laindon "A real garden city without the aid of philanthropists and on a perfectly sound basis, is likely to be created."[3]

The plotland sites have several common characteristics. They are invariably on marginal land. The inland Essex sites are all south of an invisible line across the county separating the more easily worked soil from the heavy clay, known to farmers as three-horse land, which went out of cultivation earliest in the agricultural depression. Other plotlands grew up on vulnerable coastal sites, of which the best-known were Jaywick Sands and Canvey Island, or on riverside sites in the Thames Valley, also liable to inundation. Or they are on acid heathland or chalky uplands. Even Peacehaven was built on an area of the South Downs where the ancient sheep pasture had been replaced by a tough, wiry grass as a result of ploughing in the Napoleonic and subsequent wars, with the result that it was the earliest to be abandoned as grazing.

Another characteristic of all the plotland areas was that the holiday home remained in the same family and became the retirement home of the first generation. What seemed to the outside observer to be inconvenient, substandard and far from the shops, was for them loaded with memories of happy summer days when the children were small. A final common attribute was the tendency of the plotlands, unless deliberate obstacles were put in the way of the residents, to be the subject of a process of continuous upgrading over time. Extensions, the addition of bathrooms, partial or total rebuilding, the provision of mains services and the making-up of roads, are part of the continuous improvement process in any such settlement that has not been economically undermined or subjected to "Planners' blight".

The conservationist literature of the inter-war years reveals the intense horror that was felt by all "right-thinking" (i.e. privileged) people at the desecration of the landscape they saw as happening everywhere. Dean Inge, a celebrated publicist of the period, coined the phrase "Bungaloid growth", with its implication that some kind of cancer was creeping over the face of the home counties. Clough Williams-

---

[2] Ward 1990.
[3] Hardy and Ward 1984, 196.

Ellis, who later built the holiday village of Portmerrion, was the author of *England and the Octopus* (1928) and editor of the compendium *Britain and the Beast* (1937), in which Howard Marshall declared that "a gimcrack civilisation crawls like a giant slug, across the country, leaving a foul trail of slime behind it". In retrospect, it is hard not to feel that part of this disgust was ordinary misanthropy. The wrong sort of people were getting a place in the sun.

The plotlands were, of all developments, the most vulnerable. They seldom complied with the building by-laws. They could be held to be a menace to public health since, like most of the homes of the rural poor at that time, they were not connected to sewerage systems. They provided very little income for local authorities, since their rateable value was very low, and their owners were not people with an influential voice in public affairs. They looked, when raw and new, more like boom towns, pioneering in the American West or the Australian bush, than like the expected pattern of urban growth in the south-east of England.

But there is an irony in the fact that the simple life and the rural weekend also attracted the liberal intelligentsia who were the backbone of the preservationist lobby. Reginald Bray was a progressive philanthropist and member, in succession, of the London School Board and the London County Council. In 1919 he left London to administer his father's estates based on Shere in Surrey. When Dr Peter Brandon studied the estate papers he found that Bray provided sites for many of the good and the great of the 1920s and 1930s, including a majority of members of the first Labour cabinet and several crusaders for the protection of the country-side. Clough Williams-Ellis was among them, while deploring the way in which "the adventurous bungalow plants its foundations – a pink asbestos roof screaming its challenge – across a whole parish from some pleasant upland that it has light-heartedly defaced".[4]

Another of the weekend residents was Bray's fellow Harrovian, the historian G.M. Trevelyan, who lamented that "the State is Socialist enough to destroy by taxation the classes that used to preserve rural amenity, but is still too Conservative to interfere in the purposes to which land is put by speculators to whom the land is sold".[5]

Time and nature have changed the plotland sites, just as they change any raw new settlement. For example, those offending salmon-pink asbestos–cement slates have, besides proving themselves as durable as other roofing materials, attracted moss and lichen so that their present appearance is like that of Cotswold stone. At the end of the century we may smile at the way the shapers of policy took it for granted that *they* were entitled to a rural retreat, while wanting to deny on aesthetic grounds the same opportunity to people further down the hierarchy of chance and income. Patrick Abercrombie, in introducing his Greater London Plan, was careful to stress this point: "It is possible to point with horror to the jumble of shacks and bungalows on the Langdon Hill and at Pitsea. This is a narrow-minded appreciation of what was as genuine a desire as created the group of lovely gardens

---

[4] Williams-Ellis 1928, 134.
[5] Trevelyan 1937, 183.

Figure 26 *Laindon. The original plotlands dream, immortalised in television drama, but virtually all destroyed in the building of Basildon new town.* Source: Hardy and Ward 1984.

and houses at Frensham and Bramshott."[6] As so often, Abercrombie failed to go with the flow of conventional opinion; he understood the aspirations of those he was planning for, in a way that his counterparts almost never did.

The postwar planning legislation, and the fact that landowners were now subsidised for upgrading marginal farmland, effectively put an end to plotland development, and planning authorities have perceived the existing sites as among the problems they are expected to solve. Sometimes the aim has been to eliminate them totally and return the land, if not to agriculture, then to public recreational use. In most places such policies have failed and have simply resulted in patches of empty, scrubby wasteland between those plots still occupied by determined people who fought planning decisions, with the result that local authorities were overruled by central government.

In some other places the clearance policy has succeeded. At Havering Park in Essex, the Greater London Council demolished all plotland dwellings to make a country park. Nearby, in 1949, the New Town of Basildon was designated to make some kind of urban entity out of Pitsea and Laindon, where, by the end of World War II there was a settled population of about 25 000 on 75 miles (120 km) of grass-track roads, mostly unsewered and with standpipes for water supply (Figure 26). More recently the Essex County Council eliminated another scattered plotland area to make the new residential town of South Woodham Ferrers. In

---

[6] Abercrombie 1945, 98, para. 235.

many other parts of the South East, planning authorities have tried to freeze development by refusing all applications for planning permission improvements and upgrading, refusals which have often been reversed on appeal.

Attitudes towards the plotlands have changed over the years. They began as a blot on the landscape. Then they were seen as odd, curious and vaguely interesting or quaint. After that, inevitably, they were perceived as a precious aspect of our heritage. At Basildon, one of the few remaining bungalows called The Haven at Dunton Hills became a plotland museum. At Dungeness in Kent, a plotland site was designated as a Conservation Area in order to preserve it from redevelopment. In the effort to save a plotland site in Swansea Bay from a bid to redevelop it, the local authority similarly designated it as a Conservation Area in 1990, on the grounds that the site was "arcadian". On other sites, even the antiquated railway carriages that the first settlers bought for £15 each, including delivery by horse-drawn transport to the site, have become precious for the railway antiquarians.

But the last word on the significance of the plotland era comes from Dr Anthony King, in his monumental, global history of the bungalow as a building type. He observes that:

A combination of cheap land and transport, prefabricated materials, and the owners' labour and skills had given back, to the ordinary people of the land, the opportunity denied to them for over two hundred years, an opportunity which, at the time, was still available to almost half of the world's non-industrialised population: the freedom of a man to build his own house. It was a freedom that was to be very short-lived.[7]

How right he was.

---

[7]   King 1984, 175.

CHAPTER 5

LAND SETTLEMENT: THE FAILED ALTERNATIVE

When Ebenezer Howard wrote the Introduction to his book, presenting his famous diagram of The Three Magnets, he put his emphasis, not on the overcrowding of the cities, but on the depopulation of the countryside. He, and the figures he quoted, were overwhelmingly concerned with the question he quoted from Sir John Gorst of "how to back the tide of migration of the people into the towns, and to get them back upon the land".

His was one of half-a-dozen books from the same decade, all of them attracting a great deal of attention in addressing this issue. The first was *In Darkest England, and the Way Out* (1890), in which William Booth of the Salvation Army advocated (and introduced) rural colonies to prepare the urban unemployed for a new life in Britain's overseas possessions. The second was William Morris's *News From Nowhere* (1890), describing a post-industrial post-urban Britain of the twenty-first century. The third was Robert Blatchford's *Merrie England* (1893), which advocated a revival of small-scale horticulture by ex-urbanites; it sold almost a million copies before the end of the century. The fourth was Leo Tolstoy's *The Kingdom of God is Within You* (1894), instantly translated by his disciples to urge readers to live a peasant life on the land. The fifth was Howard's neatly invented combination of proposals from 1898, and the sixth was Peter Kropotkin's *Fields, Factories and Workshops*, published in 1898 as a collection of the articles he had been publishing all through the decade on integrating decentralised industry with intensive agriculture, and combining brain work and manual work.

This proliferating literature led to a series of experimental communities, many of them in the same districts where land was cheap because of agricultural collapse.[1] Among the colonies inspired by Tolstoy, a sole survivor celebrates its centenary in 1998. This is Whiteway in Gloucestershire where the original settlers ceremonially burned the title deeds to the land to ensure that it was common to all. Inevitably this was challenged in the courts many decades later, and to the delight of most, the Chief Land Registry Tribunal ruled in 1955 that the colonists as a whole were the licensees of their land, with their monthly meeting held regularly since the formation of the colony, the licensor.[2]

---

[1] Hardy 1979.
[2] Thacker 1993, 143.

Whiteway survived through the tacit abandonment of the founders' faith in communal living, and in all these "Land Colonies" there were inevitable disagreements between rival versions of the Good Life. The Clousden Hill Free Communist and Co-operative Colony was established on a 20-acre (8 ha) farm north of Newcastle-upon-Tyne in 1895.[3] Its founder was a Czech tailor, Frank Kapper, and this was a time when there was intense interest in the North East in the potentialities of communal intensive horticulture. Kropotkin's articles in *The Nineteenth Century* and his, as yet untranslated, book from 1892, *La Conquête du Pain*, "drew attention to the potential of applying artificial heating systems, greenhouses (or 'glass culture') and new fertilisers, to land cultivation". *Freedom* serialised an English translation, *The Conquest of Bread*, in 1893–94, and coincidentally, the issue of reorganising agriculture within a democratic framework was raised in the North East by the Co-operative Movement. In May 1894 the annual "parliament" of the retail and producer co-operatives – the Co-operative Congress – met at Sunderland, and on the agenda was a special paper dealing with "Co-operative Agriculture". This attracted the attention of anarchists who tended to have a soft spot for the co-operatives, seeing them as in essence voluntary, open associations of consumers and producers, successfully eliminating the private profit motive but hamstrung by bureaucratic leadership.[4]

One of the debaters at this Congress was a Londoner, John C. Kenworthy, who was urging delegates to support "voluntary co-operation on the land" rather than farms that just chanced to be owned by retail co-operative societies. He set up a fringe meeting on this theme, where Frank Kapper met the provider of capital for the purchase of Clousden Hill. This was William Key, who had been a seaman for 12 years, a miner for 8, and a publican and part-time insurance agent for another 12, a background as improbable as that of Ebenezer Howard. Key and Kapper, anxious to do the right thing, wrote to Kropotkin (then living at Bromley in Kent, where English Heritage erected a Blue Plaque in 1989 to commemorate his residence), asking him to act as treasurer.

Kropotkin replied that, "I am the least appropriate person, as I was never able to keep accounts of my own earnings and spendings". He did, however, offer valuable advice for this and other community ventures:

> Kropotkin warned about dangers posed by the venture by insufficient funds, influxes of too many new-comers at times of prosperity in the colony, any failure to appreciate the need for hard work, and frustration that might arise out of the limited social life in small colonies . . . And he suggested that successful communities should be avoided in favour of combined efforts by independent families. Colonies should also reject internal authority structures.[5]

He raised an issue of enormous relevance for community ventures: the situation of women. It was important, he reminded them,

---

[3] Hardy 1979, 180; Marsh 1982, 100.
[4] Todd 1986, 8.
[5] Todd 1986, 19.

"to do all possible for reducing household work to the lowest minimum . . . In most communities this point was awfully neglected. The women and girls remained in the new society as they were in the old – slaves of the community: Arrangements to reduce as much as possible the amount of work which women spend in the rearing-up of children, as well as in household work are, in my opinion, as essential to the success of the community as the proper arrangements of the fields, the greenhouses, and the agricultural machinery. Even more. But while every community dreams of having the most perfect agricultural or industrial machinery, it seldom pays attention to the squandering of the forces of the house slave, the women."[6]

The Clousden Hill venture aroused enormous interest and a stream of visitors and suffered an embarrassment of recruits, all of them anxious to change the rules: "day after day was spent in framing sets of rules", wrote one colonist. Nor did its eventual failure discourage other experiments. Each of those inspirational "Back to the land" books of the 1890s gave rise to a crop of horticultural experiments. Blatchford's *Merrie England* induced a Manchester printer, Thomas Smith, to change his occupation and to move with his family to 11 acres (4.5 ha) of heavy clay at Mayland, near Althorne, and to advertise for fellow colonists. It took him a long time to succeed, and to learn from experience:

> The most profitable produce at Mayland was tomatoes and other salad vegetables, and the earlier the crop the higher the price. Smith therefore steadily moved to culti-vation under glass, producing strawberries, lettuce, tomatoes and even melons – all crops whose wholesale price was good even with small quantities. Gradually he acquired the knowledge and skill to make his holding into a thriving business. Later he published handbooks on intensive cultivation, although the picture they give of a scientifically managed market garden with intensive manuring, acres of cold frame, carefully regulated cloches and a large packing shed is perhaps not the pastoral image Smith or others had before them when setting off back to the land.[7]

Smith's rare success attracted the attention of an American philanthropist, Joseph Fels, founder of the Fels–Naptha Soap Company. George Lansbury, the Labour politician, had drawn him in to collaboration with the Boards of Poor Law Guardians in London, under the provisions of the Unemployed Workmen's Act, which granted government money to various local unemployment committees to enable them to find work. With Fels' aid, Lansbury set up "labour colonies" at Hollesley Bay in Suffolk and at Laindon in Essex, where the work of 200 men, Lansbury claimed, "turned what was derelict land into orchards and gardens". He and Fels were preparing further schemes when a change of government early in 1906 brought a new president, John Burns, to the Local Government Board, who forbade the investment of public money in schemes for the resettlement of unemployed men on the land. Undeterred, Fels went ahead with the purchase of the 600-acre (250 ha) Nipsells Farm at Mayland, close to Thomas Smith's land, with the aim of providing "a long-term opportunity" rather than "short-term relief", with Smith as manager.

---

6   Todd 1986, 110.
7   Marsh 1982, 116.

In 1912 a well-known advocate of smallholdings, F.E. Green, reported that this venture had not succeeded, "but then who could expect to find a French garden situated four and a half miles from a railway station a commercial success?" He found that most of the smallholders were deeply in debt to Mr Fels:

> Many of these settlers came from Woolwich and other urban districts, and yet one cannot lay the blame altogether on the unfitness of the men. In my opinion, Mayland should never have been cut into five-acre fruit farms, but rather into thirty- or forty-acre stock-raising holdings. A life which presents to the townsman six months of digging heavy, dirty land, unrelieved by any other winter occupation, is a sore test to the most ardent of earth lovers.[8]

Green identified the difficulty that besets every small grower, whether individual or collective: that of effective marketing. He observed:

> I was shown how the system of co-operative distribution in sending away the produce of all in bulk to market had been perfected, so I was told, "up to the last button"; but what was the use of that when the produce was sent to Covent Garden on the chance of what it might fetch? . . . In many instances produce hardly covered the cost of carriage . . . Co-operation merely perfected a method for making the fortunes of Covent Garden salesmen. This might have been avoided had co-operative distributors come to the rescue of producers.[9]

The link between Ebenezer Howard's initiatives and the range of experiments in land settlement was made by Thomas Adams, who was not only the first full-time officer of the Garden City Association, and the first general manager of Letchworth Garden City, but even the first president of the Town Planning Institute. He and the novelist Rider Haggard had organised a conference on rural depopulation, and Adams published the findings as *Garden Cities and Agriculture*,[10] while Haggard in the same year reported to the government on the land colonies established by the Salvation Army.[11]

It was World War I that changed the aspiration for resettling the land from an experiment conducted by Tolstoyans, anarchists, simple-lifers and the Salvation Army, into a minor aspect of government policy. County councils had been empowered by the Small Holdings and Allotments Act of 1908 to acquire land and build buildings with exchequer funds and rent holdings of from 1 to 50 acres (0.4 to 20 ha). They were also enabled to promote and capitalise the organisation of co-operative societies among smallholders. Ninety years later there are English counties where, because of that Act, the county council is the largest single landowner. Some counties have waiting lists of applicants for smallholdings, and as vacancies occur, face the dilemma of whether to create a new tenancy or to rent the holding to neighbouring tenants who claim that their 50 acres are too small for financial viability in the modern agricultural world. Other county councils, to raise revenue, sell holdings either to tenants or on the open market.

---

[8]   Green 1912, 258.
[9]   Green 1912, 261.
[10]   Adams 1905.
[11]   Haggard 1905.

The impetus for settling families on the land after World War I was, however, an aspect of that urban dream from the 1890s, that was reshaped for postwar aspirations. A Land Settlement (Facilities) Act was passed in 1919 and its provisions ended in 1926. These included farm colonies with central farms attached, profit-sharing farms and co-operative marketing.

> Of the millions demobilized only forty-nine thousand applied for small holdings and only about a third of these had received statutory small holdings by December 1924 . . . As a result of this war-induced land settlement, statutory small holdings were more than doubled in number, and the number of houses on small holding projects quadrupled . . . By 1924–25 the thirty thousand holdings of the combined pre-war and post-war estates had about eighty-two hundred houses upon them . . . A further thirty-six hundred of councils' holdings were "partially equipped", usually with buildings only. The remaining 60 per cent, or eighteen thousand, were bare land holdings without houses and buildings and were supplied close to the applicant's established residence.[12]

In Scotland the issue of resettling ex-servicemen on the land was given additional impetus by historical circumstances. The "Clearances" of the crofters in the Highlands and Islands had left appalling grievances; these had not been rectified by the Crofting Act of 1886 which controlled rents and gave security of tenure, but did not reinstate the descendants of the evicted. In the period leading up to World War I there had been a series of widely publicised land raids.[13] "Recruitment propaganda for the Great War promised men who enlisted voluntarily that they would get land on their return. Those who fought and survived and wanted holdings were widely considered to deserve them."[14]

About 90 per cent of the land acquired in Scotland for land settlement between 1919 and 1930 was in the crofting counties, constituting about 60 per cent of the 2536 holdings created, largely in an attempt to fulfil "the long-standing cultural and political aspirations of the crofting population".[15] In England, as provision for ex-service families dwindled, the Religious Society of Friends, popularly known as the Quakers, sought to find ways of alleviating the hardships endured by unemployed miners. By 1935 it had persuaded the government to match the funds from other sources and initiate the Land Settlement Association (LSA), specifically for the relief on unemployment, and based on collective marketing for the families involved.[16]

As sites around the country were bought, a characteristic LSA landscape emerged, recognisable even today in places disposed of by the LSA long before its final closure, and after generations of cosmetic change (Figures 27 and 28). There was a small home farm, usually the original farmstead, occupied by the supervisor or advisers, with central buildings for the grading and packing of produce, and

---

[12] Smith 1946, 109.
[13] Craig 1990.
[14] Lenenan 1989, 20.
[15] Gold and Gold 1982, 129.
[16] McCready 1974.

Figure 27 *Land Settlement – Before. A Land Settlement Association house of the 1930s, in its pristine form just after construction.* Source: LSA.

Figure 28 *Land Settlement – After. The same house in the 1990s, transformed by conversions and additions into an executive des. res.; a perfect illustration of the capacity of a house to grow and change.* Source: Brian Goodey.

beyond it about 40 holdings of around 4 to 8 acres, depending on the original assumptions about horticulture or stock-rearing as the basic activity. The tenants' houses, each with a small front garden, were built when possible on existing roads. Where necessary, new access roads were developed on a grid-iron layout. Close to the dwellings were glasshouses, pig sheds and chicken-houses, followed by a patch for fruit and vegetable cultivation, and beyond that an area designed to be ploughed and harvested together with neighbouring plots. Sometimes there was also a large-scale orchard.

It was a landscape that resembled, if anything, that of the plotlands, which in some districts, like the Selsey Peninsula in West Sussex, were contiguous. World War II both denied the LSA its triumphs and spared it the problems of its failures. For when established horticulturists were already bankrupt, it was not likely that unemployed families would do any better. Those settlers who had failed to adapt to the growers' life moved back to their home regions, where suddenly, through the magic of war, mining and heavy industry had again become important. Food production too became a national imperative and the LSA fell under the direct control of the Ministry of Agriculture.

Postwar policy was to restrict applicants for holdings to people with proven farming experience and with access to enough capital to maintain the holder and the family until they were self-supporting. In the 1960s the Ministry appointed a committee chaired by Professor M.J. Wise to report both on smallholdings established by county councils and on those of the LSA. He concluded that the concept of the Association's estates as "the first step in the farming ladder" was no longer relevant, and that its role as an experimenter in agricultural co-operation had not been fulfilled, since its board was appointed by government and not by the tenants, and since tenants were constrained by involuntary contractual obligation.[17]

Meanwhile, the pattern of retailing in Britain was rapidly changing. The concept of local greengrocers and fruiterers buying in the nearest wholesale market, dominated by Covent Garden in London, was being replaced by direct purchasing by multiple chain-stores making their own deals with suppliers for high street supermarkets and out-of-town hypermarkets, with a high degree of prepackaging and standardisation.

The LSA took the best available advice and contracted with the large multiple stores to supply a small range of salad crops in vast quantities. By the early 1970s earnings were well above the average agricultural wage, but the late 1970s brought hard times to tenants. The Ministry's decision to close down the LSA was announced as Parliament went into recess in December 1982. The decision covered the 10 remaining estates, comprising 3900 acres (1580 ha), with 530 tenants who were to be allowed to purchase their holdings at half the current market price. In what was known as the *annus mirabilis* of British farming, since farmers' incomes rose by 40 per cent, it was found that up to a quarter of LSA tenants were in receipt of the social security Family Income Supplement.[18]

[17] Wise 1967.
[18] Ward 1983.

There were messy lawsuits, successful in getting a large out-of-court settlement from the Ministry. One Suffolk estate at Newbourn formed a new co-operative to recapture their market but were defeated by cheaper imports, and by 1994 it was reported that,

> Large areas, where once a thriving community of families worked the land, now look like a bomb site. Acres of glasshouses stand idle. Clearing the glass costs £10,000 an acre, and with 25 to 30 acres of glass on the LSA site, this means a £250,000-plus bill. Growers want to sell up, but the council, in pursuit of its planning policy has banned any new building, and wants the site to retain its horticultural character. One grower for 17 years remarked that "There's just no future in horticulture; it is obsolete and we can no longer make a living at it. They want our holdings to be left as museum pieces, but without the curator's wages."[19]

This is the sad and muted end of the longest largest-scale venture in getting back to the land in Britain. Howard himself, at the beginning of the century, had remarked that "While the age we live in is the age of the great closely compacted, overcrowded city, there are already signs, for those who can read them, of a coming change so great and so momentous that the twentieth century will be known as the period of the great exodus."[20]

He was prophetic, but the exodus was not to the hard-won pastures of the promised land. It was to an ampler style of urban commuting.

[19]   Ward 1994.
[20]   Howard 1904.

CHAPTER 6

A CONTINENTAL INTERLUDE

Howard's ideas soon were exported across the Channel, to diffuse amazingly quickly across the European mainland; within less than two decades, Garden Cities were being built in France, in Germany, in Russia, and many other countries. And the movement went on reverberating: after World War II, Europe began to reconstruct its major cities around the notions of what were variously described as new towns and satellite cities. Many of these efforts themselves belong to the select Pantheon of classics of the modern planning movement. The only problem is that hardly one of them corresponds to Howard's basic idea; on the contrary, virtually all of them, however sophisticated, are garden suburbs. The failure of comprehension is so complete, so universal, that the British observer is bound to spend a few moments asking why.

SORIA'S LINEAR CITY

In point of chronology, the first of these efforts actually predated Howard, even if its realisation came a decade after him. In 1882 the Spanish engineer Arturo Soria y Mata wrote an article advocating the notion of *La Ciudad Lineal*; 10 years later he published a more detailed proposal. His notion was that a tramway (or light rail) line system running radially out from a city could be used as the basis of a linear Garden City.[1] Like Howard, he practised what he preached: the first section of his planned 30-mile (48-km) city was actually begun in 1894 and completed in 1904, running for 3 miles (5 km) circumferentially, between two major radial highways east of Madrid. On either side of the tramway – originally horse-drawn, and electrified only in 1909 – villas were laid out on superblocks measuring approximately 200 metres in depth and with 80- or 100-metre frontages. But this linear city was a pure commuter suburb, to be developed as a commercial speculation; developed by the Compania Madrilena de Urbanizacion, it was a Spanish version of an idea that had already been widely used in the United States.[2] Today the linear city survives, crossing the main highway from the airport, the tram replaced by a metro; one station is thoughtfully named after the originator.

---

[1] Soria y Puġ 1968, 35, 43.
[2] Soria y Puġ 1968, 44–49, 52.

SELLIER'S CITÉS-JARDINS

In France, the first important interpretation of Howard's ideas, Georges Benoit-Levy's *Le Cité Jardin*, confused garden city and garden suburb, a problem that would prove endemic.[3] Twelve years after that, in 1916, an important experiment in garden city construction began around Paris, in the Office Public des Habitations à Bon Marché du Département de la Seine; between 1916 and 1939, its director, Henri Sellier, planned and built 16 *cités-jardins* around Paris. Sellier knew exactly what he was doing, for he took his architects to visit Unwin in 1919, and used Unwin's text as a kind of bible.[4] But everywhere, the result was Hampstead rather than Letchworth: pure garden suburbs, just beyond the city limits, connected to commuter train lines. At the start, they resembled Hampstead: they were small, with between 1000 and 5500 units; they were built on cheap urban-fringe land; they had densities that were low for Paris, 40–60 persons to the acre (95–150/ha); and they had plentiful open space. Later, rising land costs brought higher densities – 80–105 per acre (200–260/ha) – and five-storey apartment blocks, though still with generous open space.[5]

THE GARTENSTÄDTE: DRESDEN, FRANKFURT, BERLIN

The Germans were even more faithful, forming their own German equivalent of the Garden City Association; as their leader Hans Kampffmeyer said in 1908, they wanted a German Letchworth.[6] But their garden city at Hellerau, started in 1908, was and is a mixture: like early Letchworth it was inspired by the Life Reform Movement, with the Deutsche Werkstätte für Handbaukunst, and a Society for Applied Rythmics; Heinrich Tressenow's housing is clearly part of the Unwin–Parker tradition; but Hellerau remains a garden suburb at the end of a tram line, a mere 5 miles (8 km) outside Dresden.

Then, after the turmoil of the end of World War I and the attempted communist coup of 1919, the German movement took a different direction. In Frankfurt am Main, the Social Democrats took control under mayor Ludwig Landmann; in 1925 they brought in the architect–planner Ernst May, to develop housing for land that the city had bought in the surrounding countryside.[7] Like Sellier, May had worked with Unwin, and maintained close contact with him. He started by thinking of pure garden cities, 15–20 miles (24–32 km) from the city, separated from it by a wide green belt. That proved politically impossible, so May developed satellite cities of municipal housing (*Trabantenstädte*), separated from the city by only a narrow green belt, or "people's park", dependent on it for jobs and for all but immediate

---

[3] Batchelor 1969 199.
[4] Read 1978, 349–350; Swenarton 1985, 54.
[5] Evenson 1979, 223–226; Read 1978, 350–351.
[6] Kampffmeyer 1908, 595.
[7] Yago 1984, 87–88, 94, 98–99.

local shopping needs, and therefore linked to it by public transportation.[8] The programme consisted of a mere 15 000 houses, built between 1925 and 1933; and much consisted of small penny-pocket developments. Even the larger and better-known *Trabanten*, which were planned in a linear belt along the valley of the river Nidda north-west of the city, are small: 1441 dwellings at Praunheim, 1220 at Römerstadt.[9] They were (and are) uncompromisingly in the modernist idiom, with their long terraces of flat-roofed houses carrying roof gardens; but they faithfully follow the Unwin prescription, in being small single-family homes each with its own tiny garden, aligned to catch the sun.

Also in the Unwin tradition is the use of the valley as a natural green belt, carrying – just as in Howard's original diagrams – intensive agricultural and open-space functions like allotments, sports grounds, commercial garden plots, gardening schools for young people, even a fairground.[10] And, even in their latter-day manifestations – submerged in a much larger postwar development, brutally dissected by urban motorways – these tiny developments retain much of their original quality; matured by 70 years of tree growth, they show just how good the modern movement could be in the right hands: Unwin was surely just old and wrong when, in the 1930s, he made himself thoroughly unpopular by holding out against modern architecture. In truth, the garden suburb could wear more than one set of architectural clothes, and May's design is as good in its way as were Unwin's and Parker's in theirs. However, this is the Unwin of Hampstead or Ealing, not the Unwin of Letchworth; from the start, these were pure satellites or garden suburbs, connected to the city centre only 4 miles (6 km) distant, by a tram service which has now been upgraded to a light-rail U-Bahn.

The other great manifestation, in those late 1920s years before the curtain fell on the Weimar Republic, was in Berlin. Here the city architect–planner, Martin Wagner, was co-ordinating a much larger housing programme, albeit equally modernist in concept; Berlin in those years was the world's most advanced city in cultural terms, and the building programme certainly expressed that fact. There is, however, a subtle difference as against May's programme in Frankfurt, so subtle that it is not always easy to grasp: Wagner rejected satellites in favour of the *Siedlung*, a concept originally developed in model industrial settlements in the Ruhr coalfield, and closely related to industrial settlements like Bournville and Port Sunlight. In it, houses were grouped around a factory, but with no physical separation from the city, and indeed with close links to it.[11] At the *Grosssiedlung* of Siemensstadt, developed by the Siemens company in the north-west sector of the city between 1929 and 1931, the visitor arrives by U-Bahn, a short ride from the city centre. And this announces itself as an urban development, more like a London County Council inner-London housing scheme of the 1950s than a garden suburb: the leaders of the German architectural world of that time – Scharoun, Bartning, Häring, Gropius – have designed not single-family homes but four- and

---

[8]   Fehl 1983, 188–190.
[9]   Gallion and Eisner 1963, 104.
[10] Fehl 1983, 191.
[11] Uhlig 1977, 56.

five-storey apartment blocks. But they are set in a huge garden space, which – ironically – now so envelops them that pilgrims find the buildings difficult to photograph.[12] An apartment garden suburb might seem a contradiction in terms, but Siemensstadt shows it to be possible.

It is above all restful and relaxing, a quality it shares with the best work of Unwin and Parker, and also with two other outstanding examples of Berlin urban design of this period: Onkel-Toms-Hütte (1926–27), at Zehlendorf in the south-west, and Britz (1925–27), in the south. They were developed by Gehag, a huge housing agency formed by merging trade-union-based building societies with the Berlin Social Housing Society: exactly the kind of agency that Howard wanted to build his garden city.[13] Both were pure garden suburbs on extensions of the Berlin U-Bahn, Hampstead-style. Onkel-Toms-Hütte is called a forest-settlement (*Waldsiedlung*), and indeed a huge canopy of trees covers the two- and three-storey modernist row houses, mostly by Bruno Taut and Hugo Häring.[14] Britz, designed by Martin Wagner himself in association with Bruno Taut, has two- and three-storey terraces grouped around the central *Hufeneisensiedlung*, a huge four-storey horseshoe around a lake.[15] All these developments have an almost magical quality, aided by an extraordinary level of maintenance that makes them seem almost new; but all are pure garden suburbs that do not even make an attempt to separate themselves from the city; their coherence, which is very real in all three cases, is an internal one, in that the street pattern very clearly relates to the U-Bahn stations that mark the entry and exit points. This is most beautifully seen in Onkel-Toms-Hütte, where the commuter exits from the station through a shopping gallery into the suburb; an almost perfectly rational and efficient arrangement.

But garden cities they are definitely not. This is understandable, perhaps, in Frankfurt; here, like Barry Parker in Manchester's Wythenshawe, May was adding a satellite development to a medium-sized city of about half a million, where a full-fledged garden city solution might have been inappropriate. But in Greater Berlin Wagner was planning the second greatest metropolis in Europe, with four million people, fully comparable to London – or so one might have thought. There was one critical difference: as Abercrombie had realised on visiting the city a decade earlier, Berlin was an extraordinarily dense and compact city of the traditional European kind. (Indeed it still is; today's visitor is bemused to find Tegel airport in the equivalent of Kilburn.) In that kind of city, particularly because local politics entered in because there was all too little money, the Weimar planners did not see the need for garden city solutions.[16]

---

[12] Rave and Knöfel 1968, 193.
[13] Lane 1968, 104.
[14] Rave and Knöfel 1968, 146.
[15] Rave and Knöfel 1968, 79.
[16] Hartmann 1976, 44.

Scandinavia Takes the Lead

Soon after Abercrombie's 1944 Plan prescribed a green belt and new town solution for Greater London, Copenhagen and Stockholm devised quite different planned solutions. In 1948 Copenhagen produced its Finger Plan. The spatial scale, as earlier in Frankfurt, was much smaller: just over one million people, about one-eighth the size of London. But the problem was similar: it was to de-congest a city that had grown densely, in a traditional radial-concentric fashion, and had reached a critical point in its development. By this time the city had spread such that the outer underground terminals were about 45 minutes from the centre, about the same in time terms as the outer tube terminals in London. So Copenhagen could have introduced an Abercrombie-style solution, but the city planners decided instead to encourage outward growth along selected suburban railway lines, thus extending the 45-minute zone further out (Figure 29). Between these axes or "fingers" of development, open-space wedges would tend to preserve themselves by reason of poorer accessibility; an argument that Soria had used in Madrid 66 years earlier.

It worked; and when the time came to revise the plan, in the 1960s, the planners decided to repeat the prescription in a slightly different form. The city's growth had by then reached another threshold, at about 1.5 million, the long-term figure in the 1948 Finger Plan, and was expected to grow to 2.5 million by the end of the century. It was no longer sufficient to improve accessibility to the centre; as in London, jobs must – indeed, would – decentralise. But the Copenhagen planners provided for this by a further extension of the Finger Plan principle: they proposed new "city sections", in effect large satellite towns of about a quarter-million people (equivalent to a major provincial city in Denmark), each with its own industrial zones and major centre, along extensions of certain fingers. Thus many of the new residents would find work near home; but with central-area jobs, a high-speed rail service would be available.

There was a major debate, not about the principle, but about which fingers should carry the bulk of the development; finally, it was resolved to concentrate the early stages of growth along the fingers running westwards towards the town of Roskilde and south-westwards towards the town of Køge, both medium-sized country towns about 20 miles (32 km) from the centre of Copenhagen, with a new centre around a station on the western finger at Tåstrup, 12 miles (19 km) from the centre of the city. In a 1973 revision the plan evolved further, with a concentration along two major transportation corridors, one running south from the narrow sea crossing to Sweden at Helsingør (Elsinore) towards Germany, the other running east–west from Copenhagen towards the western region of Denmark. Major employment centres would be planned along them, with regional sub-centres at certain intersections with the original fingers, three west of the city, a fourth to the south-west. And growth in the attractive northern open zone has continued to be restricted.

Stockholm in the early postwar years was (and indeed still is) even smaller than Copenhagen: while Copenhagen reached 1.1 million people by 1945 and 1.5 million by 1960, Stockholm reached only 850 000 in 1945 and 1.2 million by 1960.

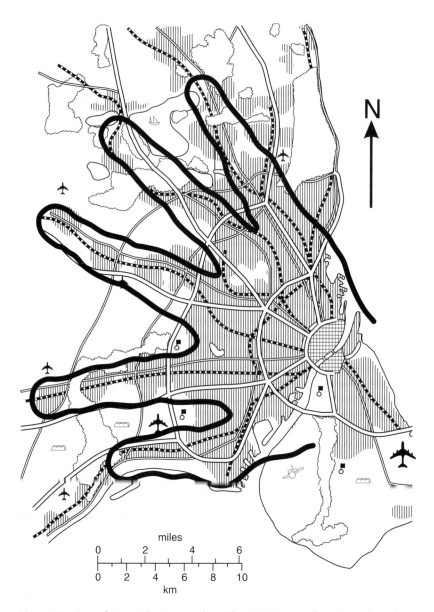

Figure 29 *Copenhagen: The Finger Plan. The 1948 Finger Plan, Copenhagen's alternative answer to the problem of accommodating metropolitan growth: instead of self-contained new towns, continuous corridors of growth.* Source: Denmark. Egnsplankontoret 1947.

Just as in Copenhagen, by the mid-1940s Stockholm had grown along tramway lines to an average distance of about 8 miles (13 km) from the centre, but was almost at the practicable limits of the system. So here, too, though a small city by European standards, the 1952 General Plan for the city, produced by Sven Markelius and Göran Sidenbladh, called for a new underground railway system, with lines radiating from a city-centre interchange station, and with stations at roughly half-mile (1 km) intervals (Figure 30). New suburban satellite units would be deliberately planned around these stations, with local pyramids of density: the highest densities would be immediately around the stations, within easy walking distance; slightly more distant areas, reached by walking or bus, would have lower densities. Shopping and other services would also be concentrated around the stations, on a hierarchical principle; each group of approximately five suburban satellites would have one major centre and four smaller district centres. These local "C" centres would serve 10 000–15 000 people, mainly within walking distance of a station, while the sub-regional "B" centres would serve the entire group, with 15 000–30 000 people within walking distance and another 50 000–100 000 served by underground, feeder bus or private car. Thus, in each case, because of the pyramids of density, local access to the shops would be maximised. Finally, each unit (and also the entire group of units) would be physically separated by local green belts, which would help give them a definite identity – both attached to the city by the new Tunnelbana, but also separate from it.

The plan was faithfully implemented. By the mid-1960s the Tunnelbana was a 40-mile (65-km) network, bringing all the new suburban areas within a 40-minute ride of the city centre. The new satellite groupings were following the extension of each line: first the Vällingby group in the west, then the Farsta group in the south, then the Skärholmen group in the south-west (Figure 31). But there was one big unexpected problem: car ownership rose far faster than the planners extended. So the Tunnelbana had to be supplemented by high-capacity arterial highways, which were upgraded to motorway standard in the late 1960s and early 1970s, while parking provision at the "B" centres had to be sharply increased. The biggest impact, though, came on commuting patterns. The plan had assumed a rather tidy pattern: half of the residents would find jobs locally, half would commute out – mainly to central Stockholm, by the underground – while half the local workforce would commute in, again by public transport. But, because of the rising car ownership, it did not work out like that at all: far fewer worked locally, typically only about 15 per cent, and far more made complex cross-city commutes which were relatively difficult to make by underground but much easier by car, especially after the opening of a major ring highway through the western suburbs in the mid-1960s. On top of that, in the early 1970s a huge flat-building programme at last ended a 20-year period of housing shortage and produced a surplus of homes; and, to the dismay of many planners, increasing numbers of households decided that they did not want to live in traditional co-operative apartments, opting instead for an Anglo-Saxon single-family-home lifestyle on the urban periphery.

As the size of the developed area grew, a new scale of thinking was necessary. So in 1973, and again in 1978, the city and the new Stockholm County Council, a creation of the 1960s, began to develop plans which incorporated not merely the

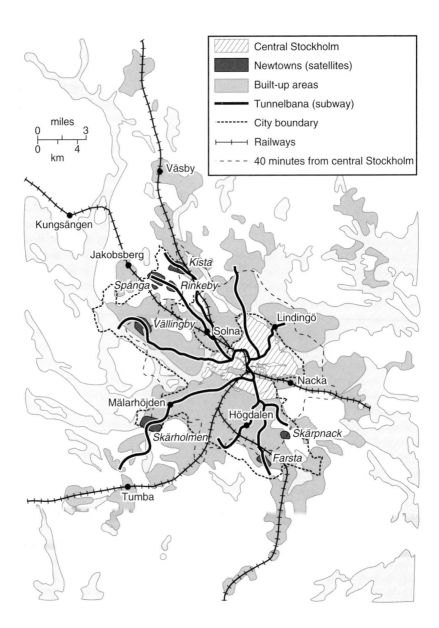

Figure 30 Stockholm: The Markelius Plan. The 1952 Stockholm Plan, from Sven Markelius and Göran Sidenbladh, established clusters of satellite towns around underground rail extensions, each with a pyramid of density: higher around the station and shops, lower toward the edge. Source: Hall 1992.

Figure 31 *Vällingby. The Markelius principle of central transport and services, and the density pyramid, in the first and most celebrated of the Stockholm satellites.* Source: Peter Hall photograph.

city, but also adjacent suburban and rural areas up to 20 miles (32 km) in all directions. Research suggested that though space-using types of industry would decentralise to peripheral locations, increasing numbers of jobs in the advanced services would still be located in or near the centre. Coupled with the outward movement of people, in search of more space and able to pay for their own homes, this implied both a big extension of the total developed area, and an increasing demand for long-distance commuter journeys to the centre; and, since central Stockholm could not absorb many additional cars, major investment in rail transport became a priority. During the 1970s, the final extensions of the Tunnelbana were completed, serving a major new development for 32 000 people in the Jarvafältet area, north-west of the city. But that was the effective limit: as in London, traditional underground rail transport would not effectively serve areas more than about 12 miles (19 km) from the centre, so that growth outside these limits had to be based on faster longer-distance main-line commuter services on the Swedish Railways system. New suburbs on this system would not concentrate so closely around the stations, but take a more dispersed form, following feeder bus routes. Increasingly, during the 1980s, the city region began to take a star-shaped form, stretching along main transport corridors westwards to Södertäjle, northwards to Arlanda airport and southwards to Tungelsta, with groups of neighbourhoods separated by belts of open land containing the major national highways. Thus the urban structure has become more discontinuous, with axial developments for many kilometres along the main highways and rail lines: a startling contrast to the much more compact patterns of the 1950s and 1960s.

PARIS 1965: HOWARD MEETS HAUSSMANN

In 1961, having resolved the Algerian crisis, Charles de Gaulle decided to resolve another equally intractable problem: how to cope with the growth of the Ile-de-France region. He toured the region by helicopter, demanding that someone "put a little order into all that", meaning the vast unplanned structure of the suburbs below.[17] The problem, as he saw it, was twofold: first, to accommodate a projected population growth for the region from 9 to 14 million between 1965 and the end of the century; second, to rectify the chronic lack of investment and planning in the suburban sprawl below. To help him, he imported his right-hand man from Algeria, Paul Delouvrier, giving him the title first of Délégué Général du District de la Région de Paris, then – from 1966 – Préfet de la Région Parisienne. A bureaucrat, not a professional planner at all, Delouvrier proved himself a master-planner in the mould of Haussmann. On appointment he was 47, three years older than Haussmann had been when Napoleon III appointed him Préfet of Paris. He later said that Haussmann had seventeen years to change Paris; he had a mere seven. But his achievement was on a similar titanic scale: just as the city of Paris to this day is essentially Haussmann's creation, so the structure of the wider region is Delouvrier's.

Delouvrier produced his master plan, the Schéma Directeur de la Région Parisienne, in 1965. To make up for a huge accumulated backlog of investment, the plan called for the creation of no less than eight new cities, strung out along two parallel axes on either side of the Seine, east and west of Paris: the first, 55 miles (90 km) long, south of the river from Mélun to Mantes, the second, 45 miles (70 km) long, north of it from Meaux to Pontoise. These new units, with populations of up to one million, would together nearly double the size of the developed area over a 35-year period to the end of the century (Figure 32).

But these *Cités Nouvelles* were in some sense a misnomer of the same kind that recalled Henri Sellier's *Villes Nouvelles*, 40 years earlier: they were not new towns on the British model, a solution the plan specifically rejected, but integral extensions of the existing agglomeration. Indeed, they were to be connected to that agglomeration by 540 miles (870 km) of new highways and 156 miles (251 km) of an entirely new regional express rail (RER) system, the first section of which opened in 1971. Also part of the plan, and inherited from a plan of 1960 which was scrapped, was the very large-scale, and hugely expensive, reconstruction of selected existing centres within the agglomeration, such as La Défense-Nanterre on the west side, then already started, St Denis in the north, Bobigny in the north-east, Créteil in the south-east, Choissy-le-Roi/Rungis (the site of the new markets of Les Halles, close to Orly airport) in the south, and Versailles in the south-west. Both the new cities and the suburban "revertebration" had a common objective: to break the monopoly concentration of economic and social and cultural life at the centre, by developing a number of urban counter-magnets.

---

[17] Anon. 1995, n.p.

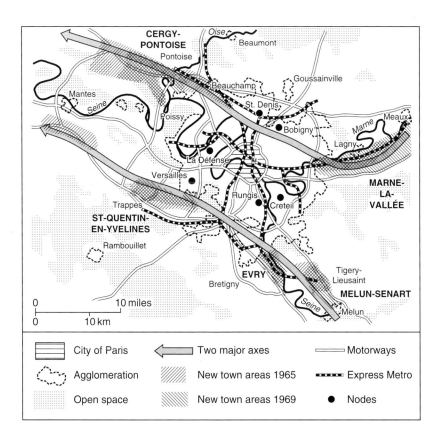

Figure 32 Paris: The 1965 Strategy. As modified in 1969, the Schéma Directeur developed five large new cities along two parallel axes, linked to the centre by a new express urban rail system and to each other by motorways. Source: Hall 1992.

Events intervened to reshape the plan: a falling birth rate caused the regional end-century projected population to be cut from 14 to 10–11 million, only marginally larger than the total achieved by 1978. (In the 1990s revision of the plan, projected growth was still minimal, from 10.3 million in 1990 to only 10.8 million by 2015.) In 1969, the number of new towns was cut to five: Cergy–Pontoise and Marne-la-Vallée on the northern axis, and St Quentin-en-Yvelines, Evry (Figure 33) and Melun–Senart on the southern axis. By 1990 they housed 617 000 people, five times the mid-1960s figure. The originally planned RER network was essentially complete by the mid-1990s; but construction of the ambitious circumferential motorway network had been delayed by environmental problems, resulting in a decision to build the western sector of the A86, the middle ring, as a deep-level toll tunnel. Shopping and public services developed remarkably in the new towns, but there remained an east–west imbalance, as commercial development showed a preference for the giant La Défense scheme and the two western new towns, Cergy–Pontoise and St Quentin-en-Yvelines; on

Figure 33 *Evry. One of the five new cities on the southern axis of the 1965 plan, with high-density housing around the town centre.* Source: Peter Hall photograph.

the east side, Marne-la-Vallée remained more dependent on public sector employment until the opening of Disneyland Paris, which has helped rectify this problem.

CONCLUSION: THE EUROPEAN TRADITION

The conclusion is clear: very consistently, mainland Europe either failed to understand Howard's argument, or wilfully misinterpreted it. From the earliest days of the movement down to the 1960s and 1970s, whether in France or Germany or Scandinavia, garden cities and satellite towns have meant new developments which are physical extensions of existing metropolitan agglomerations, with either minimal physical separation or none at all, and with associated investment in urban transit systems to increase accessibility and reduce travel times for commuters. Early on, in both France and Germany, the stress was on pure garden suburbs with either minimal associated employment, or employment in a single industrial plant – in effect, industrial model towns on the Bournville–Port Sunlight model. Later, after World War II and above all in the 1960s, there was a much greater stress on developing independent sources of employment both in manufacturing and in services, and – associated with the latter – a deliberate attempt to build up selected centres as counter-magnets to the attraction of the metropolitan centre. But such attempts were clearly limited by the fact that the strategies simultaneously increased access to that centre. The notion of deliberately creating more or less self-contained new towns or new town clusters, distant from the metropolis and outside its daily sphere of influence, was clearly foreign to the continental mind: an English notion that did not travel well across the Channel.

PART II

THE COMING CENTURY

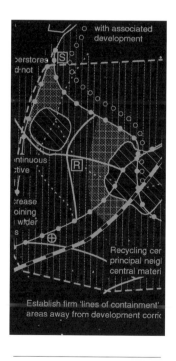

CHAPTER 7

THEN AND NOW

One hundred years later on, the question must be: how much of Howard's message is relevant to our world, and how much needs to be bent and adapted to meet the needs of the twenty-first century? For the world has changed out of recognition since he wrote. He could not have easily foreseen the passage of two world wars and a major intervening depression, the long postwar boom, global energy crises, the globalisation of the economy, or the structural shift towards an economy based not on making goods but on processing information. He could never have dreamt of a society where one in three of Britain's children, projected to rise soon to nearly one in two, receive a higher education. He could never have imagined a world in which two-thirds of us own our own houses and our own cars, where almost all of us have telephones and colour television sets, where ordinary folk regularly fly in airplanes to the most exotic parts of the world.

Garden cities, remember, were conceived in a quite different world: a much poorer and in many ways a simpler world. They were answers, brilliant answers, to the major problems of ordinary people in the 1890s: the desperate plight of poor rural labourers, living in wretched hovels, without work and without hope, ignorant and in fear; and the equally dire fate of those who had made the long trek to the cities, and who there found themselves effectively imprisoned in over-crowded tenements from where they sought miserably-paid casual work.

THE THREE MAGNETS REVISITED

Garden cities were the third magnet that could transform the lives of both these groups. And today, the diagram of the three magnets appears not merely archaic in execution, but positively perverse in its account of the good and the bad sides of town and country. As the mathematicians would say, today all the signs seem to be reversed. It is not the countryside that is losing people and jobs, it is the city. It is not town-and-country that is offered as a possible dream; it is everyone's ambition, and so successful a prescription has it proved that the countryside is resisting further encroachments.

So the odd fact is that we can redraw the diagram for the very different world of the 1990s; but it comes out very differently (Figure 34). The town now has some of the less attractive characteristics of the Victorian countryside: its factories have closed because of competition from cheaper sources abroad, and it offers a contrast: excellent jobs in the new global informational services economy for

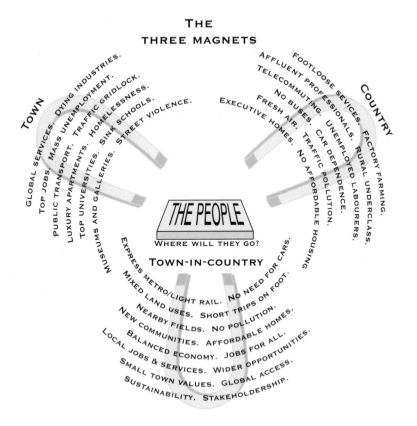

The
THREE MAGNETS

TOWN
GLOBAL SERVICES. DYING INDUSTRIES. TOP JOBS. MASS UNEMPLOYMENT. PUBLIC TRANSPORT. TRAFFIC GRIDLOCK. LUXURY APARTMENTS. HOMELESSNESS. TOP UNIVERSITIES. SINK SCHOOLS. MUSEUMS AND GALLERIES. STREET VIOLENCE.

COUNTRY
FOOTLOOSE SERVICES. FACTORY FARMING. AFFLUENT PROFESSIONALS. RURAL UNDERCLASS. TELECOMMUTING. UNEMPLOYED LABOURERS. NO BUSES. CAR DEPENDENCE. FRESH AIR. TRAFFIC POLLUTION. EXECUTIVE HOMES. NO AFFORDABLE HOUSING.

THE PEOPLE
WHERE WILL THEY GO?

TOWN-IN-COUNTRY
EXPRESS METRO/LIGHT RAIL. NO NEED FOR CARS. MIXED LAND USES. SHORT TRIPS ON FOOT. NEARBY FIELDS. NO POLLUTION. NEW COMMUNITIES. AFFORDABLE HOMES. BALANCED ECONOMY. JOBS FOR ALL. LOCAL JOBS & SERVICES. WIDER OPPORTUNITIES. SMALL TOWN VALUES. GLOBAL ACCESS. SUSTAINABILITY. STAKEHOLDERSHIP.

Figure 34 The Three Magnets, 1998. Howard's famous statement of advantages and disadvantages, rephrased for the conditions of the 1990s. The town has been sanitised and the country has been given urban technology, but both still suffer problems, and, still, towns set in the country offer an optimal lifestyle. Source: Peter Hall.

some, long-term unemployment for those who lack the education and the skills to compete for the new jobs. And hence, the town now displays large and increasing differences in income, moving us back towards that Victorian world of inequality. Also, though the Victorian peasouper fogs have gone, they have been replaced by photochemical smog from traffic. There is gridlock in the streets and poisonous concentrations of pollutants above the streets. And this is ironic, because the town still preserves a good level of access by public transport – above all in London, where the underground has created a much better quality public transport system, at cheaper real cost, than in Howard's day. Finally, the town is still perceived by many as a deprived and crime-ridden and problematic place; many of its schools in particular are seen as doing a bad job, so that parents tend to move out in search of better education for their children.

The countryside too has been transformed. Electricity and the motor car make it easy to get to all kinds of opportunities. The telephone and fax and e-mail make it easy to communicate. Central heating, gas-fired in the towns, oil-fired in the open countryside, gives a level of comfort that Earls and Dukes would have envied a century ago. Television and the video recorder, and now the Internet bring into the home a cornucopia of entertainment and instruction, inconceivable in Howard's time. But life, including access to opportunities, is too often completely dependent on the private car. And the bottom 25 per cent are just as much left out as they always were.

The remarkable fact is that over the intervening century, we have turned the English countryside into a version of Howard's town–country on a vast scale. Part of it of course was conscious: we created the Mark One new towns on his model in the 1940s, the planned town expansions in the 1950s and early 1960s, and then the Mark Two new towns in the late 1960s and 1970s. But much of it was spontaneous and market-driven, in the form of millions of people pulling up sticks and moving from the towns to the country, where they either bought and fixed up the old farm cottages of the Victorian labouring poor, turning them from hovels into luxury residences, or alternatively moved into new executive homes on the outskirts of the same villages. The pubs they visit at the weekend, once sordid places with at best a game of skittles, are now luxury eating and drinking places; the farm labourers have disappeared, making way for the middle classes. It evidently works, and people presumably like the life, or they surely would never have opted for it. They see it as offering the combination of winning features that Howard so presciently foresaw in 1898. They have been both pushed by what they saw as the bad things about living in the cities – the familiar litany, however exaggerated, of bad schools, crime, danger, poor environment – and pulled by their relative absence in the small towns and the villages.

This is confirmed by a most thorough study of attitudes to housing in England in the survey by Hedges and Clemens for the Department of the Environment, based on a survey of 3285 households in 1992. It is clear that people's satisfaction increases as one goes from big cities through smaller places to rural areas, though the authors warn that there may be some bias in the intermediate categories: twice as many people in rural areas are "very satisfied" with their location as those in urban/city centres. The same responses are clear in reply to the question about overall satisfaction with housing. So there is an inverse relationship between satisfaction and population density: at densities of less than 5 persons/ha, 68 per cent are "very" satisfied, while at 40 persons/ha and above the figure falls to 37 per cent. And another question shows that 10 times more people preferred houses to flats. A total of 64 per cent said a garden is "very" important, and 19 per cent that it is "fairly" important.[1] The movement out of dense cities, and into suburbs and country towns and villages, is fuelled by deep popular preferences. Call them prejudices if you like, but to ignore them is politically suicidal.

---

[1] Breheny 1997, 213–214.

THE REPOPULATION OF THE COUNTRYSIDE

As a result, over the last half-century we have witnessed the re-population of the English countryside. Already by the 1960s, the map of population change was the exact reverse of the equivalent map of the 1890s; the counties and the districts that were then suffering the biggest population losses have become the areas with the biggest gains. And the rings of population growth around the major cities have steadily moved further out each decade, until by the 1980s and 1990s they had begun to coalesce in the most rural fringe areas. Tony Champion's detailed analysis of migration patterns shows that at the beginning of the 1990s, "even in a year of relatively low migration from metropolitan to non-metropolitan Britain, the 'counterurbanisation cascade' seems to be a dominant feature of population distribution in Britain";[2] in other words, we are moving to smaller and smaller places (Table 1). "At all levels and in almost all cases the net result of population changes between types of places is a shift down the hierarchy;"[3] this is true not only for the whole country but also for each of four sample regions, though it is most marked in southern England.

It is not quite what Howard imagined, and we can well believe that, if he could by some miracle be brought back to speak to us today, he would not approve of it at all. He would have been both startled and saddened that the sharpest gains have been not in "districts with new towns", which exhibited only modest gains by migration, but in the most rural areas of all, those termed "remote urban/rural", "remote rural", and "most remote rural". In particular, he would not at all have liked the fact that the process has been socially exclusive, virtually shutting out lower-income groups who remained trapped in ghettoised public housing estates in the cities. And he would not have condoned the dependence on the car, or the need to travel long distances to work or to entertainment.

He might have said that we never really experimented with his Social City solution, within which everyone could walk in a few minutes to work, to shop, to school, to parkland and to open countryside; and also, through the mechanism of a light rail system linking each community with all the others, had access to a much wider range of opportunities – for varied employment, for meeting people, for entertainment. We did produce something that superficially resembles it, as we can see from an airplane today: the system of towns against a background of open countryside, which Raymond Unwin so enthusiastically proposed. But many of them are too large to be true walk-to-work places, and they lack efficient public transport to link them up; they fail at both the local and the regional scale.

There are two other features of this new geography that need emphasising, and Howard might find them the most surprising of all. The first is that, sub-optimal or not, people evidently value this life so highly that they will fight to the death to preserve it. And the enemies they see, ironically, are the city-dwellers who have the same values and the same hopes, and who want to come and join them. In July

---

[2] Champion and Atkins 1996, 21.
[3] Champion and Atkins 1996, 26.

Table 1 *Population changes arising from within-Britain migration, 1990–91, by district types.*

District type	Population 1991	Net migration 1990–91	
		Number	%
Metropolitan Britain	*19 030 230*	*–85 379*	*–0.45*
1 Inner London	2 504 451	–31 009	–1.24
2 Outer London	4 175 248	–21 159	–0.51
3 Principal Metropolitan Cities	3 922 670	–26 311	–0.67
4 Other Metropolitan Districts	8 427 861	–6 900	–0.08
Non-Metropolitan Britain	*35 858 614*	*85 379*	*0.24*
5 Large Non-Metropolitan Cities	3 493 284	–14 040	–0.40
6 Small Non-Metropolitan Cities	1 861 351	–7 812	–0.42
7 Industrial Districts	7 475 515	7 194	0.10
8 Districts with New Towns	2 838 258	2 627	0.09
9 Resort, Port and Retirement	3 591 972	17 736	0.49
10 Urban/Rural Mixed	7 918 701	19 537	0.25
11 Remote Urban/Rural	2 302 925	13 665	0.59
12 Remote Rural	1 645 330	10 022	0.61
13 Most Remote Rural	4 731 278	36 450	0.77

Source: Champion and Atkins 1996, from Census 1991.

1997 the *Sunday Times* carried a feature that Nick Raynsford, the Minister for Housing and Planning, was admitting the case that major development would be needed in the countryside to house a projected 4.4 million additional households expected between 1991 and 2016. One self-appointed spokesperson was reported as saying that one million people would march on Parliament if this were to happen. Exaggerated this may be, but it gives a sense of the NIMBY sentiments that grip these new ruralites. In the event, a quarter of a million people attended a rally on 1 March 1998; there was a suggestion in the media that it was organised by a well-heeled field sports lobby, and that the other issues had been added to disguise that fact.

THE PARADOX OF RURAL LAND VALUE

The other feature of this new geography is that, in truth, the economic case for preserving this countryside appears about as problematic as in Howard's time. Now, as then, and this is perhaps the only constancy, much of this land is effectively lying idle. Then, landowners were going bankrupt; now, the European Commission rewards them handsomely by setting their land aside and watching it grow no crop at all. In 1995, an astonishing total of 544 900 hectares, 5.9 per cent, of the entire farmed area of England was set aside in this way, effectively growing nothing. In the South East region 148 097 hectares, 8.9 per cent of the total farm area, was idle. Essex had 10.8 per cent of its farmed area set aside; Hampshire, 8.6 per cent; Oxfordshire, 10.0 per cent; and Bedfordshire, 12.2 per cent (Table 2).

Table 2 EU Set-Aside land, 1995.

	Hectares	% farm area
Bedfordshire	10 870	12.2
Berkshire	6 806	9.5
Buckinghamshire	11 084	8.9
East Sussex	6 461	5.7
Essex	28 573	10.8
Greater London	1 106	8.0
Hampshire	19 457	8.6
Hertfordshire	11 920	11.5
Isle of Wight	1 466	5.7
Kent	19 095	7.7
Oxfordshire	20 409	10.0
Surrey	2 976	4.6
West Sussex	7 876	6.4
South East Region	*148 097*	*8.9*
England	*544 005*	*5.9*

Source: Ministry of Agriculture, Fisheries and Food 1996.

The result ought to be that its economic value is near-zero (or, more accurately, the value of what Brussels is willing to pay in subsidy). In fact, it is more complex than that: agricultural land is indeed not worth much, but if anyone can obtain planning permission to develop it, then its value will increase a hundredfold; and this would be pure development gain, no longer subject to any capture on the part of the community that created the value by its decision.

There is one exception to this, and an important one: it is called the Pointe Gourde principle, and it says that the community shall not be required to pay any development value as the result of a planning scheme. A planning scheme, in this legal language, is a large development, such as a new town, that has basically created the gain in development value. There is thus an implicit distinction between what are often called the normal processes of development, where the planning authority provides for the extension or infilling or rounding-off of a town or village, and a substantial new settlement that is effectively the result of a public action. This is the last surviving element of what was originally a very ambitious attempt – some would say far too ambitious – to capture all development value in the original 1947 Town and Country Planning Act. The logic behind that attempt was that the 1947 Act effectively nationalised the right to develop land, and provided for compensation to be paid to those who could show they had lost development rights. The Conservative government in 1954 first repealed most of the provision, and then in 1959 went further by providing that in cases of compulsory purchase the public authority should pay market value – with this most conspicuous exception.

The result, ever since, has been a rather extraordinary situation: thousands of people flock into the countryside each year, yet rural planning authorities consistently resist providing the land for them; in this situation, anyone who can get planning permission stands to make a killing, thus increasing the pressure to make planning applications and to appeal against a refusal of permission to develop. Further, though there has been a great deal of debate on this point over the years, it is now generally agreed that this restriction on the amount of development land does raise the price of that land and thus of the housing that is built on it.

One recent definitive statement on the topic, by Gerald Eve and Partners in 1992, concluded interestingly that in order to reduce the cost of land appreciably it would be necessary to release very large amounts, thus in effect compromising the entire system of planning and development control we built up in 1947.[4] Even more recently, there has been another comprehensive exercise by Glen Bramley and his colleagues. Like Gerald Eve, they show that if we released large amounts of land the effect would not be as great as we might think, because of the "implementation gap" before the new policy had an impact on the round.[5] The price effect might average between 4.5 and 7.3 per cent over a period; this would range from around 9 per cent in growth districts (8.9 per cent in Berkshire), down to 2 per cent in declining industrial areas. In terms of output of homes, even the most drastic version of land release would make an average output difference of about 2.7 per cent and a maximum of 5.5 per cent;[6] the authors comment "It is questionable whether such a gain would be worth the environmental and political costs".[7] The effects on output would, however, mean greater geographical concentration in certain areas, increasing pressure for new settlements and local controversies about large-scale developments.[8]

Another rather surprising conclusion is that if house prices rose or fell, it would not make much difference to the density of new developments – certainly less than has been found for America. This is partly because British densities are much higher than American ones, so that densities cannot be easily boosted, but also because the British planning system restricts variations in density.[9] Bramley and his colleagues conclude that the planning system actually restricts adjustments to market forces, but that "If the housing environments produced are broadly what people want, then this may not be a bad price to pay for local democratic control over the process of land development".[10]

The authors also conclude that Alan Evans may be right when he concludes that planning policy in the South East has concentrated high-density development in or adjacent to urban areas, because of an excessive protection of rural areas.[11] "If the regulators have been captured," they say, "it is more by the rural lobby and suburban residents than by the housebuilding industry."[12] The spread of owner-

---

[4] GB Department of the Environment 1992b.
[5] Bramley et al. 1995, 165.
[6] Bramley et al. 1995, 153–154.
[7] Bramley et al. 1995, 154.
[8] Bramley et al. 1995, 165.
[9] Bramley et al. 1995, 184.
[10] Bramley et al. 1995, 189.
[11] Evans 1991.
[12] Bramley et al. 1995, 221.

occupation has worsened this problem by giving voters an acute equity stake in their local environment. They think that there is a need for more realistic targets that allow for windfalls and avoiding delay, perhaps by setting targets that are upper limits.[13]

So, Bramley and his colleagues conclude, "planning does affect the housing market, raising house prices and densities and reducing supply responsiveness" but "An absence of planning, or a very liberal planning policy on housing land release, would not eliminate these problems, or even drastically reduce them" and some of the failings could be significantly ameliorated by a modified planning style, for instance in relation to targets in structure and local plans.[14] And, in any case, "planning is almost certainly here to stay" because so many people have a stake in it.[15] They think that planning agreements could play a bigger role in securing land and especially subsidy for social housing.[16]

The implications are ironic: if a latter-day Howard sought to build a garden city, however well sited and well designed, he would almost certainly be refused planning permission and the Secretary of State for the Environment would uphold the decision on appeal – if only because it was contrary to the local development plan, which under the 1990 Town and Country Planning Act must be the "chief material consideration" in determining the application. There is no apparent way that a single iconoclastic individual like Howard could build a Letchworth or a Welwyn Garden City, unless through some public agency with exceptional powers, like a new town development corporation. Significantly, when the Town and Country Planning Association tried in the 1970s to initiate an experimental self-build community, they had to look first to Milton Keynes and then to Telford new town. And that is the ultimate irony: Howard, who so distrusted government as an agent to build new towns, would today find himself totally dependent on its active co-operation. We return to that specific dilemma in Chapter 11.

That is a fact of life: since 1947 we have had a comprehensive land use planning system, something that did not even exist when the Reith Committee reached its deliberations, and clearly any proposal for a new town would have to fit within the structure of development plans and planning permissions. We have also moved, in that half century, from a world in which the great majority of housing was built by public sector agencies to be let on subsidised rents to lower-income people, to a world in which the great majority of new housing is built for sale on mortgage. And – as seen at the end of Chapter 3 – principles developed for a public service ethos will not transfer over to a world in which the profit ethos rules: some new way must be found of capturing a share of land value for the community whose actions created it, while allowing market mechanisms to continue operating smoothly.

There is, however, an additional problem: if it was difficult politically to launch new towns in the late 1940s, it is doubly or trebly difficult now. NIMBYism, a

13 Bramley et al. 1995, 221–222.
14 Bramley et al. 1995, 235.
15 Bramley et al. 1995, 235.
16 Bramley et al. 1995, 235.

movement that existed then even if no one had given a name to it, is now immeasurably stronger, and it has extended outward from an original prejudice against public sector building for lower-income groups, to an opposition to any form of development: as one commentator has coined it, BANANA, Build Absolutely Nothing Anywhere Near Anything. That applies equally to the most conventional speculative builder's estate of junior executive detached housing, as to more experimental communitarian building which reflects the original co-operative spirit of Howard's tract.

THE CHALLENGE OF 1998: SUSTAINABLE GROWTH

The central challenge, for the first 20 years of the coming century and probably beyond that, is to deal with an unprecedented growth of new households. The 1992-based revised household projections from the Department of the Environment, published in late 1995, suggest that between 1991 and 2016, just 25 years, we may have to find housing for some 4.4 million additional households in England alone (Figure 35): 3.5 million by 2011 alone. This is the largest upward jump ever recorded, a 70 per cent increase on the previous 25-year forecast of 2.6 million, which has formed the basis for housing allocations in recent Regional Planning Guidance. Even those more modest figures were hotly disputed by some local authorities and environmental groups, particularly in the South East. The growth in households will happen not because of population growth, which will remain reasonably modest, but because of changes in the composition of that population and also because of social changes – many more young people leaving home for higher education or first jobs, many more divorces and separations, more old people living longer but eventually getting widowed.

Of the 3.5 million forecast increase in households to 2011, no less than 2.76 million – 79 per cent – will be one-person households, and more than half of these will be never-married people living alone. The younger members of this group may well be satisfied to live in rented high-density apartments close to shops and entertainment; this might require massive conversion and densification of existing suburbs, which might not prove politically acceptable. In any case, as they age and as their incomes rise, even small households may well spread themselves into more generous apartments, and if they can afford the price the market will meet their demand.

To make it more daunting, two in five of the additional households, 1.73 million out of the 4.4 million, are expected to be in the South East: almost one in two if we add areas on the fringe of the South East, which happen to form part of the belt of really rapid population growth in Britain at the present time, a belt that stretches up from Bournemouth, through Swindon, to Milton Keynes and Northampton, and then snakes along the A14 to Cambridge and on to Ipswich. In other words, the additional pressures will be greatest where there is already the most pressure, and the most controversy about new development, plus some areas to the west and north of them, in counties like the former county of Avon, and in Somerset (Figure 36).

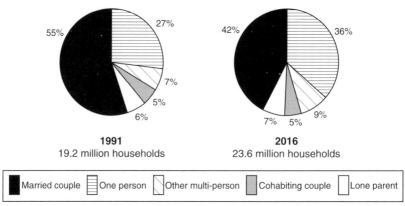

Number of households by household type: England: 1991 and 2016

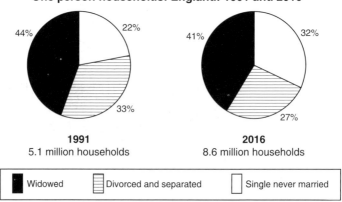

One person households: England: 1991 and 2016

Source: DOE, Projections of Households in England to 2016, HMSO, 1995

Figure 35 *Household Projections. The projections for the years 1991–2016 show 4.4 million additional households in England alone, 1.73 million of them in the South East, and no less than 79 per cent of them one-person households – the product of more young people leaving parental homes, rising divorce and separation rates, and more old people surviving their partners for longer periods.* Source: GB Secretary of State for the Environment 1996.

Projections are of course apt to prove wrong. But they can go wrong either way, and this particular projection is more likely to be revised upwards than downwards, because it does not include accumulated shortages and the development of obsolescence in the existing stock. Alan Hooper points out that in the 1980s we were building only half the new homes that we completed in the 1960s, yet there is no evidence that demand is weakening; the shortfall is now running at over 100 000 units a year. He agrees with Alan Holmans that restricting supply does not reduce the formation of households; it simply increases the number of sharing and

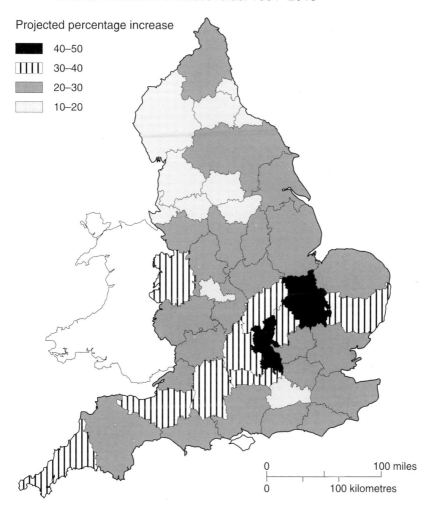

Growth in number of households: 1991–2016

Projected percentage increase

- ■ 40–50
- ⦀ 30–40
- ▨ 20–30
- ☐ 10–20

0 ——— 100 miles

0 ——— 100 kilometres

Source: DOE, Projections of Households in England to 2016, HMSO, 1995

Figure 36 *Regional Household Projections. The big increases are projected to occur in the "Sunbelt" across the borders of South East England, from Avon via Northamptonshire to Cambridgeshire. However many are housed on brownfield sites, there will be a need for extensive greenfield development here. Source: GB Secretary of State for the Environment 1996.*

concealed households, and in particular is likely to increase the need for social housing.[17] But in any case, because of the uncertainty, the right way would surely be to use a two-stage approach – both central and local government – needed to ensure adequate land banks for release to meet demand, but then to adjust the actual flow to meet changing circumstances.

Some have argued that the projections are circular: as with roads, so with housing, the more we provide, the greater the demand. But Holmans' careful analysis of housing demand and supply over the last 40 years suggests that at national level there is probably no circularity. The circularity problem arises at more local levels, where an increase in local housing supply may have a disproportionate feedback effect on in-migration and vice versa. This makes it particularly difficult for local authorities to target housing for lower-income groups; without a significant municipal programme of new homes, any supply may simply be occupied by more affluent in-migrants. This point cannot be over-emphasised: if we fail to build enough market housing, the resultant inflation of housing costs will drive ever more households to compete for a limited supply of social housing.

At the end of the day, the question is not merely Howard's old question "The People – Where Will They Go?" but also "The People – How Should They Live?" The key issue is whether we can find ways of housing the 4.4 million people in ways that are decent, acceptable to the affluent people of an advanced nation entering the twenty-first century, and also compatible with the imperative to protect our environment. The fashionable logic suggests that, by packing more people into existing urban areas, we can both reduce the need to travel – thus reducing fuel consumption and emissions – and minimise the loss of open countryside to development. The Government's PPG 13 of 1994 accepted this goal, and the 1995 Housing White Paper proposed that 50 per cent of all residential development should be on recycled urban land by 2005,[18] compared to an equivalent figure of 49 per cent in 1993.

The TCPA's own report on the problem, based on seven regional conferences in mid 1996, suggested that this was unduly optimistic; over a 25-year period, 30–40 per cent might be more realistic.[19] This was not, as some commentators commented, a "target"; it was the best estimate of what in fact would happen, based on the recent rate of actual brownfield re-use development and on the message, consistently given to us by local planners at the conferences, that these recent rates of re-use, which have reached close to 50 per cent, could simply not be sustained under the present national and local policy regime. To achieve a higher level of re-use in the future would require powerful new initiatives. Without substantial national government support, it is difficult to see how a large-scale programme of brownfield development can be sustained.

It followed that the previous government's own revised "aspirational target" of 60 per cent brownfield development, contained in its November 1996 Green Paper,

---

[17] Breheny and Hall 1996a, Chapter 2.
[18] GB Department of the Environment 1995.
[19] Breheny and Hall 1996c.

or the 75 per cent target proposed early in 1997 by the UK Round Table on Sustainable Development, were even less realistic;[20] they were frankly unachievable without a sharp deterioration in urban quality, as Ministers in the incoming Labour government have argued. That makes it the more surprising that the Labour government, having firmly reduced the brownfield target from 60 to 50 per cent in November 1997, promptly raised it back to 60 per cent – as a 10-year objective – in the February 1997 White Paper.[21] Cynics might say that these strange somersaults reflect purely political realities, not the reality of the underlying brute facts. Cynics might be right: an aspirational target (a wonderful Sir Humphreyesque phrase!) is not so much different from a 10-year target that happens conveniently to be two governments away.

And perhaps longer than that: the Llewelyn-Davies review of urban capacity studies for the UK Round Table highlights not just the funding gap – detailed in their earlier report on brownfield development in Strathclyde – but also a deep reluctance on the part of local planning authorities to move away from existing planning policies and standards on matters like densities, parking standards or amenity levels. The consultants imply that local authorities are being stubborn in clinging to established practices; but, equally, they may be seeking to protect the quality of life of their local communities, which is at least as legitimate for them as for ruralites. NIMBYism is not just a rural phenomenon; the problem now is that the real needs of people for housing will be crushed between the two millstones of urban and rural resistance to any form of change.

SOME BASIC PRINCIPLES . . .

The central challenge, then, is to reconcile the need to house the 4.4 million households decently, to good design standards, and to do so while respecting the principles of sustainable development. It may seem difficult, even impossible. In fact it is not so difficult, if we keep a clear head and start by establishing some basic principles.

In their response to the November 1996 Green Paper, the TCPA tried to do precisely that.[22] They started from the premise that *all decisions to develop land should be based on a common and agreed framework of environmental considerations*. This framework should embrace all elements of the environment – social and economic as well as environmental. The central concern should be the impact on individual human beings, both in relation to the wider human society and to the wider ecosystem to which they indissolubly belong.

In such a system, *land per se has no privileged place*. It is merely one element in the total ecosystem. True, rural land, once urbanised, is unlikely to return to rural uses, but the Planning Minister Richard Caborn has reminded us that over the last 20 years we have actually *doubled* the area of green belts, so that they now

---

[20]   GB Secretary of State for the Environment 1996; UK Round Table 1997.
[21]   GB Deputy Prime Minister 1998.
[22]   Anon. 1997b.

far exceed the area covered by housing. And loss of land in itself is not necessarily a key consideration, so long as the remaining rural land is capable of yielding equivalent amounts of renewable natural resources, and so long as the effect on the total ecosystem (e.g. in terms of photosynthesis) is not adversely affected. The definitive research on the subject, commissioned by the Department of the Environment and published together with the 1996 Green Paper, comes from Peter Bibby and John Shepherd, and has formed the basis of the Department's own calculations.[23] Their definitive conclusions are these.

First, between 1991 and 2016 some 169 000 hectares (1.3 per cent of the area of England) are projected to change from rural to urban uses. This is an area slightly larger than the county of Surrey, or the districts of East and South Cambridgeshire; not, one would have thought, very significant in the overall picture. Second, the result of this is that by 2016 about 11.9 per cent of the land area of England will be in urban uses as against 10.6 per cent in 1991, a growth of 12.2 per cent on a very modest base. Third, this projected conversion from rural to urban uses would account for some 2.3 per cent of the land area in rural uses which is "undesignated" – that is, not reserved in the form of National Parks, Areas of Outstanding Natural Beauty, Green Belts and the like. Fourth, between 1991 and 2016 households are projected to increase by 23 per cent but land in urban uses is projected to increase by only 12.2 per cent – a result of the simple fact that about half of all development has recently been in urban areas and – according to the government's own statement – will continue to be over the projected period. Fifth, these conversions are equivalent to a change from rural to urban uses of about 6800 hectares a year, about two and a half square miles, or a little bigger than the borough of Poole in Dorset; a growth rate of less than 0.5 per cent per annum; one might be tempted to say, statistically insignificant.[24] This needs to be compared with the figures reported earlier in this chapter: in 1995 there were in England 544 000 hectares of farmland set aside under the EU's agricultural policy, growing nothing: *three times the amount of land needed to accommodate all urban development over the coming quarter century.*

Now it is true that the impact will be geographically concentrated, especially in the South East: in London, but also in Berkshire, Buckinghamshire and Bedfordshire. Further, the relative rate of urban growth – expressed as a percentage of the urbanised area at the start – will be particularly high in what geographers and estate agents now call the golden belt, stretching from Cornwall and Devon through Somerset, Dorset, Wiltshire, Oxfordshire and Buckinghamshire to Northamptonshire, Cambridgeshire and Suffolk. The results are shown in Figure 36. But nowhere does this proportion rise much above 20 per cent – and high figures like that simply express the fact that we are starting from a slim base figure.

Finally, Bibby and Shepherd show that the conversions of so-called undesignated land also show great variation. Their most alarming figure is that in Greater London the urban growth just exceeds the amount of undesignated rural land; in

---

[23] GB Department of the Environment 1995.

[24] Bibby and Shepherd 1997, 117–120; GB Department of the Environment 1995, viii.

other words, we would be urbanising more land than we have available to urbanise. That is a measure of town cramming, because it must mean taking urban greenfield land that never ought to be taken.

Therefore, we argue, *the idea of a sequential test (whereby it would have to be shown that no brownfield land was available before any greenfield land was released) should be rejected*: it amounts to a reassertion of the concept of the onus of proof on the putative developer, which was suggested by the Scott Committee 55 years ago, but has never once been accepted as a principle of British town and country planning.

The most important reason for concern about rural land use is not about land in itself, but about the effect on motorised travel, which contributes disproportionately to consumption of non-renewable resources (fossil fuels) and to pollution, and which research has shown to depend on the density and form of development. But it needs stressing that these relationships are far from fully explored, and are still the subject of fierce disagreement among experts. The ECOTEC research,[25] which provided an important basis for the government's PPG 13 on Planning and Transport, shows that large urban areas are more economic in terms of energy consumption for transport than smaller towns and rural areas (though new towns emerge just below the national average); however, higher densities in the urban areas are accompanied by high proportions of low-income people dependent on public transport, and the effects are difficult to disentangle. What has not been proved is that greenfield development is intrinsically energy-profligate; as will be seen in Chapter 8, theoretical work both in Britain[26] and elsewhere[27] suggests that it is possible to design highly energy-efficient new communities based on local mixed land uses and a linear emphasis on strong public transport corridors.

The right policy, based on the principle of equal and dispassionate application of environmental criteria, is *to use principles of sustainable urban development wherever that development takes place*. If substantial amounts of brownfield land are available in urban locations with good public transport access, it will be right to develop them at relatively (but not excessively) high densities, as long as the quality of the development is always high. That, after all, has been a principle of good planning for half a century, as embodied in the classic Markelius–Sidenbladh 1952 plan for Stockholm, which was discussed in Chapter 6. However, it should be remembered that this plan was based on peripheral greenfield development; it is always easier to apply correct principles in greenfield development, because the physical constraints are fewer. There is a real danger that poorly conceived and poorly implemented town cramming policies will prove less sustainable, in every sense, than well-conceived greenfield development. High-density schemes crammed next to polluting main roads or railways, or on environmentally problematic sites, do not provide an answer, particularly when used to house the least fortunate members of society: such schemes should be rejected out of hand.

---

[25] GB Department of the Environment and Department of Transport, 1993.
[26] Breheny and Rookwood 1993, Owens 1992a, 1992b.
[27] Calthorpe 1993.

. . . AND BASIC CONSTRAINTS

In addition to those basic principles, the TCPA suggested that policy would have to be guided by certain basic – even brutal – realities.

First, *there will be no additional financial resources to revive the cities.* This has been unaffected by a change of government; previous levels of spending have continued, and will maintain urban renewal projects at their barely adequate but modest current levels. It follows that *the 60 per cent (or higher) re-use target cannot be achieved* with existing levels of funding; such levels would imply a massive diversion of resources into decontamination of polluted land, expensive infrastructure preparation, and subsidies for difficult sites.

Just what these costs might be, have begun to emerge from the TCPA's research on brownfield capacity financed by the Joseph Rowntree Foundation. This presents ten case studies. Six were commercially freestanding and did not require a government grant. Where a grant was required, it ranged from £70 000 to £115 000 per acre, or £5000–£11 000 per unit. The average level of subsidy was £100 000 per acre, or £7400 per unit.[28] The importance of high values is illustrated by two sites in the South East – in West London and in Hertfordshire. They had the highest levels of abnormal development costs in the sample, £330 000 and £400 000 per acre respectively, or 17 and 15 per cent of total development costs – but, because of high site values, they did not need grant funding. The unimpaired site values were approximately £670 000 and £1 000 000 per acre respectively.[29]

With one exception, the non-subsidised sites all had very low abnormals, ranging from 1 to 6 per cent of total development costs. In all cases where grant funding was required, the purchase price was discounted at more than 50 per cent from open market value. "Clearly," the study concludes, "at some point, the discount required to make the scheme viable will render re-use as housing uneconomic without some form of public sector intervention. As a derelict site, there may be greater value attached, for example, to continued industrial use or as a scrapyard."[30] The question is whether in practice government might be willing to provide these subsidies. It could be done by cross-subsidies from a levy on greenfield development. But the costs involved are clearly going to be substantial.

Second, *the revival of urban living will require a broad strategy of renewal – social, economic and educational just as much as physical.* The problems of the cities are only partly physical: they include poor education (massively underlined by publication of the first primary school league tables), crime and welfare. It is for these reasons, just as much as the poor physical condition of many inner-city areas, that people are fleeing the cities. Tackling these problems will require resources and time – suggesting that the genuine revival of our cities is a very long-term project.

Third, *the economic geography of the twenty-first-century UK will continue to favour decentralised locations.* True, certain sectors – financial services, quality

---

[28] Fulford 1998, para. 1.4.4.
[29] Fulford 1998, para. 1.4.4.
[30] Fulford 1998, para. 1.1.4.

consumer and cultural services – will continue to locate in the hearts of the major cities. But most other economic activities will continue to gravitate towards locations that are either well located for the motorway and trunk road network, or to small towns and rural areas in accessible locations. It will be difficult to persuade large numbers of people back to the cities and towns if their jobs are outside them. So, while a revival of major towns and cities may slow the process down, and would be highly desirable, the process of urban decentralisation must be taken as a fact of life.

Fourth, *the majority of households will continue to prefer suburban and semi-rural areas*. The definitive research by Hedges and Clemens, reported near the beginning of this chapter, makes that clear. In the longer run, it may well be possible to change attitudes about both locations and preferred types of dwellings. This particularly applies to some of the single-person households that constitute 79 per cent of the net increase expected by 2011: students and young people, in particular, may find cosmopolitan inner-city living, with good services and entertainment, highly congenial. But they are likely to occupy existing urban stock for the most part, displacing conventional families who will move out. And research for the Housing Research Foundation, published in April 1998, suggests that many of them – especially the separated and divorced – strongly prefer houses with gardens.[31] For the conventional family units which will still represent the great majority of all households by 2011/2016, existing preferences and prejudices towards city living will prevail. That is particularly true if, as we suggest, resources are not made available to tackle the multiple problems of those cities.

Fifth, *severe fiscal restraints on the use of the car will not be introduced*. Ever since PPG 13, planning policy has recognised the overriding priority to reduce the need to travel, above all to travel in solo-occupancy cars. But this approach requires that we take steps to achieve it, most obviously through the price mechanism. Politically, as discussions of road pricing both in Britain and elsewhere have suggested, this is difficult. The TCPA welcomes the recent government decision to call upon local authorities to adopt traffic reduction strategies, and indeed believes that the transport budget should be steadily and significantly reoriented towards such strategies, including physical and fiscal restraints coupled with promotion of public transport and ride-sharing. But, realistically, the experience of places which have earlier adopted such policies – such as some German cities – suggests that at best they are likely to slow down the growth of car use.

It is true that widespread leaks, bearing all the signs of authenticity, have suggested that the Government White Paper on Integrated Transport, expected in June 1998 but now delayed, would make fairly radical proposals: a high rate of taxation on all commercial off-street parking, including supermarket parking, as well as provision to allow local councils to charge for the use of congested roadspace (road pricing) if they wish. There is said to be an ongoing battle between the Department of the Environment, Transport and the Regions on the disposition

---

[31] Housing Research Foundation 1998.

of the tax income: the Treasury, as usual, wants to claw most of it back into the general tax pot while the DETR is anxious that councils are allowed to retain it to invest in public transport and other sustainable transport measures.

The 1998 budget did nothing to resolve any of these questions; and by June, another shower of rumours was suggesting that the more radical proposals were being watered down at the Prime Minister's insistence, after focus groups had suggested that they would be exceedingly unpopular with the electorate of Middle England. What would remain was more money for investment in public transport, but without the draconian measures that might have moved large numbers of people out of their cars. So it would appear that our basic assumption will have proved right.

It follows, sixth, that *people will continue to favour car travel*. In the absence of severe fiscal measures to restrain car use or a short-term change in attitudes, cars will continue to dominate our lives. There will be particular resistance to measures to restrain car use (e.g. low or zero parking standards in new housing), particularly since this housing will have to sell in the open market, and mortgage lenders may be reluctant to finance it. This need not preclude experiments such as greater emphasis on on-street parking combined with traffic calming, as in the latest Californian examples of the "new urbanism"; it does, however, suggest that overall car dependence is unlikely to be reduced.

Seventh, *there will be local resistance to excessive intensification*. Despite the new agenda that promotes urban compaction in order to deliver sustainable development, local authorities will defend existing levels of amenity, just as they always have. Local communities will object to small-scale schemes which involve loss of garden space or additional parking and traffic, and local plans will recognise this fact. Large-scale densification will involve a degree of rebuilding that is not conceivable, given the fact that the vast majority of the housing is in the hands of owner-occupiers who have no wish for change. This will set limits on possible densification.

The TCPA response emphasised that it was at least as fully committed to policies of traffic reduction, and to policies to find more urban development land through well-conceived higher-density schemes, as any other interested group. The TCPA portfolio approach fully recognises the contribution such schemes can make; the Association was particularly supportive of the work of the Urban Villages Forum, which is developing models of sustainable urban development. But we must recognise the limits of the possible. As the TCPA warned: "The very real danger is that, by pursuing policies and targets that have no prospect of implementation, we achieve a worst-case outcome which is satisfactory to no one."[32]

The TCPA statement argued that the concept of the Sustainable Social City Region has very real merits. It totally follows the principle of urban compaction, but in a way that does not compromise the quality of urban living and working. It provides for economic activities to function efficiently but sustainably, and for ordinary people to live the lives they want to live. It is based on the principle of regional and sub-regional strategies based on rigorous and even-handed

---

[32] Anon. 1997b, 81.

environmental impact assessment; it fully embodies the principle of re-use of urban sites as a priority where appropriate, but gives the necessary flexibility for those cases where other forms of development may be more desirable and may actually deliver greater sustainability benefits. Above all, it implies a *programmed approach*, in which local authorities make clear their priorities for development at a regional and then at a local scale.

This in turn implies that top-down, regional-level principles need to be blended with bottom-up local perspectives on what is required in any given area. The resulting blend should then make sense in principle and practice. This blending, or *portfolio* approach, is central to the Social City Region concept. It consists of a carefully chosen combination of different elements – brownfield regeneration, intensification where appropriate, urban extensions where they do not violate green belt or other designations, village infillings and extensions, and new communities; all developed so as to maximise access to employment, shopping and social opportunities, especially by walking, cycling and public transport. This is the approach that will be spelt out in more detail in Chapters 8 and 9.

Building Sustainable Communities: The Key Issues

There is, therefore, a whole raft of issues that need to be resolved if we are ever again to create new communities.

The first is to deal head-on with the issue of *sustainability*, the great planning catchword of the 1990s. There is a powerful campaign to the effect that sustainability demands that we maximise building in the cities, and avoid greenfield building save as a last resort. We have just shown that this is not necessary to save land; in Chapter 8, which follows, we look at the argument in terms of other resources, particularly the energy requirements of transport and the polluting emissions that come from it. Then, in Chapter 9, we suggest how it would be possible to produce a portfolio of brownfield and greenfield development, in each region and sub-region, that would satisfy the most rigorous demands of sustainability.

The next question, logically, concerns the mechanics of development: *agencies and finance*. We have seen that, ever since Howard's day, these interrelated questions have proved some of the most intractable: we have never been able to devise a solution that successfully combined public and private agencies and financing, and we have never, from Letchworth onward, come near Howard's vision of a self-governing, self-financing commonwealth. We look at that question in two sections. First, in Chapter 10, we put forward a package approach which would secure simultaneous large-scale regeneration of brownfield land and the creation of new communities on transport corridors leading from our cities and conurbations; second, in Chapter 11, we ask how Howard's original communitarian dream could after all be realised.

Finally, since it will not go away, we come back to the political problem of NIMBYism (or BANANAism). The opponents of development, we recognise, are passionate and well organised, and, as demonstrated in the nationwide countryside protest of March 1998, part of a wider groundswell which may represent a fundamental shift in the structure of interests that underlies British democratic politics. Any solution, to be viable, must take account of such political realities, and this we try to do in our concluding Chapter 12.

CHAPTER 8

THE QUEST FOR SUSTAINABILITY

In 1998 we might say that we are fully briefed, almost over-briefed, on the subject of sustainable urban development. But there is a problem: though everyone knows that much quoted and generally accepted *general* definition of sustainability, from the Brundtland Report of 1987 – "development which meets present needs without compromising the ability of future generations to achieve their own needs and aspirations" – it is not clear how this maps into actual everyday decisions in everyday urban contexts.

Of course, the broad policy objectives are clear enough. There is fairly general agreement that we shall need a combination of policies, some concerned with individual building standards, some with transport, and some with land use. Thus we should develop building forms that conserve energy and minimise emissions of pollutants; encourage accessibility without mobility, or specifically without the need for mechanised transportation (particularly by providing for places to be reachable on foot and by bicycle); encourage public transport and discourage single-user driving; develop new forms of propulsion which are less polluting and more economical of energy than the internal combustion engine; and develop centres of activity around public transport nodes. The problem is the next stage: to translate these objectives into workable strategic frameworks and plans for real places. To discuss these topics fully would take a book in itself. What is critical for the purpose of this particular book is the land use–transportation interface, which Howard clearly thought important in 1898. But it now has a new dimension, and it poses a problem which Howard could never have imagined.

It is simply this: not only in the UK (Chapter 7), but everywhere in the advanced urban world, alike in the developed heartlands of Europe and North America and Japan and in the explosively growing cities of Asia and Latin America, growth inevitably means internal dispersal. This is a matter of sheer physical capacity: new homes for new residents can only be found at the periphery, local service jobs must follow them, and in the largest urban areas local labour forces and markets exert a pull on more basic economic activities. New and efficient factories, representing inward investment, tend to seek generous space for single-storey operation and for storage, and connections with national highway systems. Warehousing is drawn to the same locations, and to container port facilities which have been located at estuarine sites well away from the city cores. And, in the largest metropolitan centres, commercial service activities are also attracted out to lower-rent locations accessible to their suburban workforces. The most dramatic illustrations are the

Edge Cities, or New Downtowns, in the American suburbs: places like the New Jersey "Zip Strip", Tysons Corner outside Washington, DC, Arlington outside Denver, Dublin–Pleasanton–Walnut Creek outside San Francisco, or Bellevue outside Seattle.[1] But European equivalents can be found in places like Reading west of London, the new towns of the Ile-de-France region, or the E4 Stockholm–Uppsala corridor in Sweden. And a development like Shin-Yokohama, on the Tokaido Shinkansen outside Tokyo, is a Japanese version of the same phenomenon. In Eastern Asia, portents of this form of development can already be seen in the new cities of the Shenzen economic zone, or the development of the sub-centres of Jakarta, or the new commercial centre of Shanghai.

They can be regarded as a natural reaction on the part of employers and developers to the facts of life in the giant metropoles: high rents, and long and expensive commuter journeys leading to enhanced salary demands. And, in so far as they represent a kind of locational re-equilibrium, bringing the work closer to the workers, they represent at least a partial answer to some of the most pressing problems of such places. But, at least in some cases, they are an imperfect solution. The American Edge Cities, in particular, are in origin pure speculative developments, fitting into no structure of regional planning, and almost exclusively dependent on commuting by the private car. Thus Tysons Corner is not on the Washington Metro system, which terminates a couple of miles away. The European and Japanese examples are almost all better located in relation to public transport, but even so most of their workers tend to be car commuters; it has proved almost impossible to provide public transport in a context of dispersed homes and dispersed jobs. Kista, in the E4 Stockholm corridor, is centre of a new high-technology zone of growth: it lies at the outer end of the city's Tunnelbana system, but beyond it – as seen in Chapter 6 – the workers depend on their cars. Therefore, places in this outer belt tend to represent a move away from sustainable urban development rather than a move toward it.

The central dilemma for urban management is therefore this. The basic structural forces are shaping ever greater concentrations of people in urban centres for the generation and exchange of information – and, in the newly industrialising nations, for producing goods. Though much information travels instantaneously and cheaply over wires or through the ether, it also travels inside the heads of information workers, requiring face-to-face contact. Therefore, at the macro-scale, there is increasing concentration into metropolitan areas and into super-metropolitan (megalopolitan) zones of growth, with intensive personal travel both to-work (commuting) and in-work. At the meso-scale of the individual metro area, there is simultaneously a process of dispersal of homes and jobs. But this may further increase the need for travel: from suburb to centre if people disperse further and faster than their jobs, from suburb to suburb in car-dependent patterns if the jobs catch up with them. And, since these people are reasonably affluent, their lifestyles generate increasing demands for non-work journeys in leisure time, further exacerbating the problem.

---

[1] Garreau 1991; Sudjic 1992.

So, to summarise: in virtually every advanced country for which we have record, since about 1960, cities have decentralised. The evidence is now overwhelming that both population and, behind it, employment are moving out, and that this process is most marked in the largest metropolitan areas (i.e. the ones with populations of one million and more).[2] The results are complex: journeys may have shortened, but a much higher proportion are made by car, and there is also a big growth of non-work trips by car. This is particularly evident in Western Europe, though Paul Cheshire has recently suggested that the process may have partly gone into reverse in the 1980s.[3]

At first sight all this does not look much of a recipe for sustainability, particularly since research by ECOTEC[4] has told us that people who live in the countryside or in small towns use more energy to travel than people who live in London or the provincial conurbations. Michael Breheny has used the ECOTEC figures to suggest that, since 1961, as we have moved out of the cities and into the small towns, we have been moving steadily away from sustainability rather than toward it; if we had stayed where we were, he concludes, we would be using about 3 per cent less energy for travel than we are now using. Even new towns, on the basis of his data, are less sustainable than older denser cities like London – though they are near the national average, and a good deal more sustainable than many other kinds of settlement pattern. However, the effect of the out-movement on energy consumption is marginal: maybe a difference in fuel consumption of about 3 per cent.[5] And, with these figures, we now face the prospect that the entire process will rapidly intensify.

The DOE, which commissioned the ECOTEC work, used it as the basis of the celebrated PPG 13, published in 1994,[6] which calls for a combination of transport policies and land use policies in order to reduce dependence on the private car. As a result, the government has been using Regional Planning Guidance to try to cram more people into the metropolitan areas. In fact, this is far from new: it started right at the beginning of the Thatcher era, and as much as anything it is political in its motivation, with the desire not to upset all those NIMBY citizens out in the shires.

But it reached a new point of intensity in a consultation paper published late in 1996,[7] which proposed to try to put 60 per cent of all new housing into the existing built-up areas. The incoming Labour government first reduced that to 50 per cent, and then – faced with a rural backlash – announced that they would try to move up to the 60 per cent target over a 10-year period.[8] Some of us strongly question whether that would be either practically possible, or truly sustainable.[9] In some of

2 Cheshire and Hay 1989; Hall and Hay 1980; van den Berg et al. 1982.
3 Cheshire 1995.
4 GB Department of the Environment and Department of Transport 1993.
5 Breheny 1995a.
6 GB Department of the Environment and Welsh Office 1994.
7 GB Secretary of State for the Environment 1996.
8 GB Deputy Prime Minister 1998.
9 Breheny and Hall 1996c.

our looser-structured conurbations, one might achieve 60 per cent, because they contain large areas of brownfield land between the towns; they are true conurbations, in the sense that Patrick Geddes used when he coined the word 80 years ago.

London: The Challenge to Sustainability

But in London, it will be a virtual physical impossibility: the space does not exist. Here the projections suggest that 629 000 new homes should be found over a 25-year period. The Department of the Environment, Transport and the Regions (DETR), together with the Government Office for London (GOL) and the London Planning Advisory Committee (LPAC), commissioned studies by consultants, Llewelyn-Davies Planning (LDP), who have devoted huge energy and intelligence to finding how we can maximise the number of homes on London land. Their report suggests that we should try to concentrate additional housing within a 10-minute walking radius around train stations, what they call the "pedshed", so that – as far as ever possible – people will not be car-dependent.

Doubtless, there are places in these zones – especially in the run-down "shatter zones" at the edge of the commercial centres – where useful housing gains are to be had. But, to do the LDP team the credit, their report also raises real doubts. For instance, they spend a good deal of space examining the potential for what they call backland development: crudely, building on people's back gardens (Figure 37). And they conclude, in italics, *"The potential for backland development is significantly constrained by practical issues of land ownership and assembly."*[10] Crudely, how does a London borough council persuade all these proud villa owners to surrender their big gardens? Typically, the report shows, anything between 10 and 150 separate ownerships may be involved. And, "Given the degree to which most backland areas are used and enjoyed by residents, it seems unlikely that, even with a more favourable planning policy approach, any more than a small proportion will be developed in a comprehensive principle following the design principles set out above."[11] That is an understatement, since the residents will pressure their local councils not to modify the stringent policies now in force; thus the "more favourable planning policy approach" is unlikely to be forthcoming. The NIMBIES are alive and well in the suburbs as well as in the shire counties.

Rather ominously, the report goes on to suggest that "new mechanisms" may be needed to facilitate redevelopment:

> The use of compulsory purchase powers may be appropriate where the quality of the dwelling stock has declined to the point that there is a strong public interest in securing its redevelopment, but new mechanisms to facilitate privately led development also require redevelopment. These might establish legal processes

---

[10] Llewelyn-Davies Planning 1997, 44.
[11] Llewelyn-Davies Planning 1997, 44.

Figure 37 *Backland Development. Proposals by Llewelyn-Davies for achieving up to 106 000 new dwellings around London suburban train stations. But how would the residents ever agree to sacrifice their cherished gardens? And the threat of compulsory purchase would cause the ultimate NIMBY movement.* Source: Llewelyn-Davies Planning 1997.

> whereby groups of owners could propose a redevelopment scheme and secure the necessary majority of other owners to its implementation.[12]

This seems to be slightly delphic language for a process whereby a particularly rapacious property company acquires houses and runs them down Rachman-style with all-night raves and drug-dealing and a few odd roaming Rottweilers, thus terrorising or cajoling the remaining terrified owners to agree to sell out. The media might have an interesting time with this, if ever it were allowed to happen.

There is a cautionary tale here, that may conveniently have been forgotten: in the late 1970s the London Borough of Wandsworth proposed to obtain a "housing gain" by acquiring a street of Victorian houses and redeveloping the site at a

---

[12] Llewelyn-Davies Planning 1997, 45.

greater density. This caused bitter controversy; at the next local election, Labour were defeated and replaced by a Conservative council that has remained in power ever since. And, at the 1979 general election, the sitting Labour MP Hugh Jenkins was defeated by a young Conservative aspirant called David Mellor. Wandsworth never did get its housing gain.

In any case, the critical question is the bottom line. The report applies the results of the case studies London-wide, and concludes that within 10 minutes of all town centres in London, if all available sites were trawled and existing policies and standards were applied, there might be a gain of 52 000 dwellings. If we applied a site-based design approach with only one off-street parking space per dwelling, then the figure would rise to 77 000 dwellings. And if we removed altogether any requirement for off-street parking, we might gain as many as 106 000 extra dwellings. It might well be reasonable, close to London town centres, to apply a standard of one parking space per household. Indeed one might legitimately argue that no off-street parking was reasonable; many perfectly fine detached villa residences in London have none. But the bottom line is that these figures – 52 000 at the bottom end, 106 000 at the top end – compare with a total predicted housing need in London of 629 000 dwellings between 1991 and 2016: in other words, they would provide for between 8 and 17 per cent of the projected need. There are other possibilities outside the pedsheds, but the LDP report thinks that they do not amount to much. What this means is that GOL and LPAC and the boroughs will have to be looking to other sources.

One is urban greenfield: urban land that has never been built on. And, basically, this divides into two: land that is pure wasteland, which never got developed because it was seen as too difficult or too unattractive or both, and land that has been reserved as parks, playing fields and golf courses, or just as landscape areas. As to the first, the wastelands, Llewelyn-Davies did another very valuable earlier study of the biggest single source of such land in London, Thames Gateway, which contains huge tracts of desolate marshland at Barking Reach and Havering Riverside.[13] They concluded that the housing yield might be 30 600 units, out of a total of about 100 000 for the whole corridor stretching for some 50 miles from the Royal Docks and Greenwich Peninsula down to the Isle of Sheppey. This raises the provision to between 13 and 22 per cent of the forecast need.

There are undoubtedly other such sites in London, though never on that scale. Some of the most interesting and important are the railway wildernesses: areas so cut up by railways that they have proved impossible to develop over the years. The odd fact here is that, because London's railways developed in such an anarchic fashion, and because in particular there was never real liaison between the underground and what used to be British Rail, there are extraordinary places in London where railways pass over or under each other without any connection. The most spectacular of all are in west London: at Chiswick Park where the Piccadilly and the District pass over the Silverlink Metro (the North London Line); at Old Oak Common where the Central Line and the North London pass respectively

---

[13] GB Department of the Environment 1993a.

under and over the Great Western Main Line; and at Wormwood Scrubs where the Great Western passes over the West London Line, soon (if Railtrack can be persuaded to agree) to be part of an Outer Circle railway. Properly developed, these could be major transport interchanges. They are all in under-developed wildernesses, and an important part of any strategy for London should be to develop them as new town centres with associated high-density residential development. Oddly, there appears to be no active proposal to this end, though when Rosehaugh Stanhope were developing Chiswick Park they did propose to build a station as part of the job. Ironically, nearly a decade later, Chiswick Park is still lying fallow.

The other part of the urban greenfield is the parks and playing fields. Michael Breheny has pointed out that nationally, in the early 1990s we were achieving around 61 per cent of new housing within the urban envelope, but that no less than 12 per cent of this was on such urban greenfield land.[14] He says, and the TCPA generally says, that this is quite wrong and ought to stop. If there is a doubt about greenfield building, then the one place there ought to be total doubt is urban greenfield. This is a precious and irreplaceable resource that is not only used for playing football or cricket or golf, but also for walking the dog or for jogging. And it is important simply to preserve the biodiversity of our cities, and to promote photosynthesis, which becomes ever more vital as the emissions of carbon dioxide mount. And incidentally, as we build on all the places like Barking Reach, which are now performing the same function but may not do so for much longer, the preservation of the remaining urban greenfield land will become even more crucial.

So we should surely place a moratorium on further urban greenfield development, with maybe one small exception. There are some very large green areas in London which are used almost exclusively for recreation, and then only at weekends for the most part. Wormwood Scrubs in west London is one such; the Lee Valley Regional Park in north-east London is another. One heretical suggestion is that there might be a case for really high-density developments next to or around such sites, perhaps taking a strip off some of the green space in return for an equivalent rededication nearby. That would especially be the case if we could use such sites as transport interchanges. It would be worth considering – but it would need to be kept under very strict control.

The final source of housing in London is of course the re-use of sites originally developed as industrial areas or warehousing or whatever – of which Docklands is of course the model. LPAC and others are of course right in saying that many such sites arrive as windfalls; we cannot easily predict them before they happen. The problem is, however, that every housing gain of this kind is a potential job loss. The rule should surely be this: if there is no realistic probability that this land is going to be viable in its original use, then it must be legitimate to recycle it for housing. And this should be the priority over other alternative uses, like multi-screen movie villages. The same goes for older offices that can be remodelled into

---

[14] Breheny 1997, 212; Breheny and Hall 1996c, 46.

apartment blocks, on the model just set by Castrol House on the Marylebone Road. LPAC estimates that 60 000 units might be obtained in this way: a modest contribution, once again, but worth having.

This long exercise in arithmetic finally brings us to the point that there is absolutely no realistic possibility that we shall ever shoehorn more than about 300 000 of the 629 000, at best, into London. LPAC, in their latest input to the new SERPLAN draft Planning Advice,[15] at the end of 1997, are reported as suggesting they could achieve some 380 000, or 60 per cent, of the total; that is not in the realms of the possible. And the worry is that we will be shoehorning with a vengeance: that we shall be developing all sorts of inappropriate sites, bad for the people who live in them, bad above all for their children if they have them. The Llewelyn-Davies report has an appendix, featuring 48 design exercises; at least 13 of them appear to abut directly on main roads with high traffic levels and consequently high levels of noise and pollution. We must ask: are these really appropriate locations for new residential development, bearing in mind the new report that came out in January 1998 from the Department of Health Committee on the Medical Effects of Pollutants,[16] which showed that all 10 of the most grossly polluted areas were urban ones – five of them in London?

Next door to London, Hertfordshire has employed consultants to look at the possibility of densification in 15 sample areas, one of which – Potters Bar – is in fact a suburban extension of London. The results suggested that the possibilities in Victorian or Edwardian terraced areas, or interwar estates, were effectively zero; there were better prospects on postwar council estates or in areas of large detached dwellings, especially near town centres. Overall, the consultants determined that with really determined policies across the county, a total of 11 000 dwellings could be provided: a net addition of 2.5 per cent. But it is clear that the practical difficulties are daunting.[17] Some are illustrated in one of the articles, which shows a block of postwar council houses emptied out of their families and replaced by one-person households, with new blocks or garage spaces across the former gardens. The existing inhabitants, the report suggests, could be persuaded to stampede by the promise of a £15 000 profit each. But realism causes the author of the article to doubt whether it would be as simple as all that. He is a Hertfordshire planner and he thinks it would all be worth while. We would ask, worthwhile for whom?

The Hertfordshire consultants stressed problems like traffic congestion, pollution and parking. In addition, there is what could be called the Raymond Unwin paradox. As David Lock emphasised in a recent review of the problem, in Unwin's famous pamphlet *Nothing Gained by Overcrowding!*, published in 1912, he showed that as you try to drive up density the bonus is less than you expect, because of the need to provide for fixed standards of provision which are related to the number of people or the number of households.[18] Unwin's key fixed element

---

[15] SERPLAN 1998.
[16] GB Department of Health 1998.
[17] Caulton 1996; Pitt 1995.
[18] Lock 1995.

was public open space, including parks and playing fields. Now it is just possible that you can squeeze those standards: since four in five of the additional house-holds will be one-person households, and by definition there will be no children around (except perhaps children visiting divorced parents), they may not need sports fields; indeed, we may need fewer schools altogether, so maybe we can redevelop them and their playing fields for housing. These new urban bachelors may want to go for walks in parks, but we might provide those in the green belt nearby, or in linear parks within London, and build higher-density housing to face on to those areas.

It could be quite exciting in terms of design; look at how Ralph Erskine, a twentieth-century master of design, and winner of the Millennium Village design competition at Greenwich, handles high-density development against a back-ground of trees and water at Vårby Gård in Stockholm. But there is an additional problem, that Unwin in 1912 did not have to consider: the cars. These one-person households will almost all want a car each, and most will be able to afford one. As David Lock points out, quoting another study by Llewelyn-Davies Planning, you can get up to 32 habitable rooms per hectare with two-storey housing, which by the way is not far from what Ebenezer Howard was proposing for his Garden Cities; but if you want 40 rooms per hectare you will need a mixture of houses and flats, though you could still get them on traditional streets; above that, you will need low-rise blocks in shared grounds. This could be perfectly satisfactory for bachelor living, as a lot of good commercial development in California demonstrates; but, as one moves towards more bachelor units the density bonus falls, because of the need to incorporate separate services for each dwelling.

So there is a dilemma for policy – not merely in London, not merely in England, but anywhere where the new household projections show the need for a major building programme. There are lessons to be learned here from historical experience, too easily forgotten; the last time we faced a challenge like this, government also encouraged the cities to crowd people in so as to avoid pressures on the countryside, resulting in uninhabitable high-rise blocks (Figure 38). But there is a way of squaring this circle. We can learn how to achieve it both by innovations in transport, and by innovations in land use. We will discuss each in turn.

POLICY RESPONSES: TRANSPORT

European countries and cities, over the last quarter century, have taken some kind of world lead in developing sustainable responses to this challenge in terms of transport policies. During the 1970s and 1980s many cities developed new public transport systems while progressively placing curbs on the free use of the car in urban areas.[19] Public transport investments have taken four main forms:

---

[19] Banister and Hall 1995; Hall 1995.

Figure 38 *The Piggeries. One disastrous illustration of the last attempt at brownfield town cramming, in the 1960s: a high-rise block in Everton, Liverpool, uninhabitable and deserted only a few years after it opened. Unless we take great care, we could be following the same path all over again.* Source: Peter Hall photograph.

1. Extensions of existing heavy rail systems in the largest cities (Paris).
2. New heavy rail systems, generally in second-order cities (Stockholm as early as the 1950s, Amsterdam, Barcelona, Brussels, Lyon, Madrid, Marseille, Milan, Munich, Oslo, Rotterdam and Vienna).
3. Transformation of old tram systems into light rail systems, generally in third-order cities (Frankfurt, Grenoble, Hanover, Nantes, Stuttgart, Toulouse and Vienna).
4. New express rail systems (the RER in Paris, the S-Bahn systems in Frankfurt, Stuttgart and Munich, the Blue Line in Glasgow, Merseyrail in Liverpool, Thameslink in London) to connect city centres with major urban extensions and with freestanding settlements within the extended commuter area.[20]

These new systems have been supported, indeed made necessary in some cases, by the growth of the bigger European cities to levels at which major new systems became viable. However, with the exception of the express systems and some limited light rail extensions along old rights of way, in general they have been restricted to the historic densely built urban envelope. There is a good reason for this: the characteristics of the journey, including average speed and seating capacity, do not make them really suitable for longer-distance operations.

Simultaneously, during the last two decades European cities have developed three striking innovations in curbing the use of the private car. First, the pedestrianisation of central business cores, associated with special preferential access for surface public transport, or by the undergrounding of surface transport. This makes car access relatively less attractive, and access by public transport more attractive; to the extent that in the most spectacular cases, such as Munich, public transport becomes the preferred means of access. A variant, developed in Italian cities (such as Florence and Milan) during the late 1980s, consists of the complete barring of the central business district to the private car during daytime business hours. Second, the use of traffic calming techniques, generally area-wide in networks of residential streets, but in a few cases – such as the Lister Meile in Hanover – to main traffic arteries, with the aim of reducing speed and flow. And the examples quoted above demonstrate the first important general conclusion: in Europe, no country has any kind of monopoly of good practice for long.

Third, the major innovation of the 1990s: urban road pricing in the major Norwegian cities – Bergen, Trondheim and Oslo – which charge for central access. The official justification of the Oslo scheme is not to restrain traffic but to pay for major road investments (in particular, a very expensive city-centre tunnel). The proposed scheme for Stockholm, now abandoned, would have had a double objective: as well as helping to finance expensive new ring roads, it would have restrained traffic in the entire inner city.[21]

All these three kinds of scheme have been specifically urban, even inner-urban; conceptually, they assume that the problems of cities can be dealt with in isolation

---

[20] Hall 1995.
[21] Sweden 1993; Tegnér 1994.

from the wider urban context. But cities and countries that have invested most ambitiously in good quality public transport also tend to have been ones that experienced the most rapid long-term rises in car ownership; maybe one is a reaction to the other. It is perhaps significant that in most European countries during the 1980s, public transport shared in the general upward trend of passenger-kilometres travelled; Britain was an exception.[22] These countries – such as France and Germany – happen to have invested more in transport generally; there is nothing particularly virtuous about this, and it would be possible to argue that one country invests too little or that its neighbours invest too much. The question could be resolved only by a very elaborate international cost–benefit analysis, which has not been made.

However, there remains a doubt. Surely, as Newman and Kenworthy have argued in a well-known study,[23] European cities perform better than their New World equivalents. But there seems to be a doubt as to where they are going. Indeed, by trying very hard to keep their major city centres strong in every way – as centres for offices, for shopping, for entertainment – Europeans may be contributing to the problem. Their critics may be right when they say that we should encourage the outward movement of employment closer to where the people actually live – a process the British have been encouraging in the London region ever since the original Mark One new towns. And urban road pricing could actually act as an agent of this process, strengthening market trends. We have to remain agnostic on this point until we have firmer research results at the level of the entire wider city region.

NEW CONCEPTS IN PUBLIC TRANSPORT: (1) THE PARIS ORBITALE

This, however, may bring another problem in its wake. Ever since Peter Daniels' pioneering work,[24] we have known that if people and their activities decentralise, there are two contradictory effects: commuter journeys are shortened, but there is a huge transfer from public transport to the private car. Further, a recent study shows that typical metropolitan areas in Europe and in America – Paris, Frankfurt and San Francisco – have all decentralised homes and jobs, leading to a huge growth in suburb-to-suburb commuting and a corresponding shift from public transport to car.[25] The dominance of the car was particularly evident for local trips to work within the outer suburbs; car journeys absolutely dominate the trip matrix in these zones, in the European cases as in the American one. Thus, though both Paris and Frankfurt have invested massively in new public transport, they have failed to adapt transit to the pure suburb-to-suburb commute. Reducing car

---

[22] Mackett 1993.
[23] Newman and Kenworthy 1989a, 1989b.
[24] Daniels and Warnes 1980.
[25] Hall et al. 1993.

dependence in these outer suburbs, then, can be regarded as the key element of a future metropolitan transportation strategy.

That suggests one solution: develop a public transport system that could cope with dispersed suburban journeys. Planners in Paris have developed a strategy to take care of this suburb-to-suburb commute problem. ORBITALE (Organisation Régionale dans le Bassin Intérieur des Transports Annulaires Libérés d'Encombrements) is a new 175-km transit system to serve the higher-density inner suburbs, running mainly on grade-separated rights-of-way, but with some street stretches, and with 50 transfer points to the existing radial transit system, to be built at an estimated cost of 40 billion francs. All sections should be complete by the time this book is published.[26] For the outer suburbs, and in particular the five new towns at an average distance of about 15 miles (25 km) from the centre of Paris, there is a longer-term plan: LUTECE (Liaisons à Utilisation Tangentielle en Couronne Extérieure), an extension of the RER (Regional Express Rail) system to link the new towns and strategic sectors with one another.[27] ORBITALE and LUTECE are part of the 1994 regional plan for the Ile-de-France, but are not consciously designed as part of an integrated land use–transportation strategy: the land uses are mainly in place and the major emphasis over the next 20 years is on consolidation.

ORBITALE and LUTECE do not, however, represent the only possible solution. Alternatively, a city might develop fleets of deregulated minibuses running in all directions, using specially designated motorway lanes wherever congestion threatened to delay them. Canadian transport authorities have developed suburban hub-and-spoke bus systems in cities like Edmonton (Alberta), and minibuses work equally well in such a system. These might incorporate new systems of electronically guided buses, introduced during the 1990s, which give many of the ride characteristics of the best light rail but at lower cost.

Here, the American approach to the problem is interestingly different. American policy initiatives have taken three main forms. First, as in Europe, there has been investment in light rail systems in a wide range of medium-sized cities (Buffalo, Pittsburgh, Portland (Oregon), Sacramento, San Jose, Los Angeles, San Diego). Second, in some places systems management approaches have given priority to shared vehicles both on the highway and in parking lots (high-occupancy vehicle lanes, ride-sharing information systems, priority parking); at least one Edge City, Bellevue outside Seattle, has developed a comprehensive and effective package.[28] Third, telecommuting, or encouraging workers to work from their homes or from local workstations, could be an extremely effective way of reducing the demand for travel, as experiments in California and elsewhere have demonstrated.[29] Fourth, and most radical, are attempts to phase out the internal combustion engine in favour of more sustainable propulsion systems (e.g. the South Coast Air Quality Management district approach in Los Angeles).

---

[26] Direction Régionale 1990, 22–23.
[27] Institut d'Aménagement 1990, 82–83.
[28] Cervero 1985, 1989.
[29] Handy and Mokhtarian 1995; Mokhtarian 1991.

Essentially, these last two policy sets accept the fact of the dispersed automobile-oriented city but seek to change driver behaviour. In some parts of California, they may well be combined with transit investment and with transit-oriented land use policies to produce an amalgam of the European and American approaches.[30]

AN ORBITALE FOR LONDON?

London, too, needs an Orbitale, for two separate but linked reasons. The first is as a critically important transport planning device, to transfer a share of the capital's orbital movements from car to public transport. Second, as recognised in the Ile-de-France Schéma Directeur, it is needed to provide a critical structural element for strategic planning, which will permit the creation of high-intensity mixed-use nodes around the intersections between this system and the traditional radially oriented public transport of previous eras.

It appears a daunting task. Strangely, it is not that difficult and not that costly. Quite large parts of London's Orbitale already exist, the product of generous investment in previous generations. Other parts are under construction. Proposals – indeed funding – exist to link them. Compared with the costs of really major schemes, like the east–west CrossRail and other Regional Metro schemes, the London Orbitale would be a bargain-basement job.

The first key element would be RingRail, long advocated by Transport 2000 and other groups. Large parts of it already exist and are in operation: the West London Line, recently reopened to a skeletal passenger service, between Clapham Junction and Willesden Junction; the North London Line between Willesden Junction and Dalston Kingsland; the East London Line of London Underground, now under extensive reconstruction with a new interchange at Canada Water with the Jubilee Line extension; and the South London Line between Peckham Rye and Wandsworth Road. The key investment, linking these, would be the proposed East London Line extension, at the north end from its present Shoreditch terminus over the old Broad Street tracks of the North London Line to Dalston and on to Highbury and Islington, and at the south end from Surrey Quays to Nunhead. Otherwise, all that is needed is to utilise an existing connection between Wandsworth Road and Clapham Junction, plus new stations on the West London Line, actually already funded, at Chelsea Harbour, Earl's Court and Shepherd's Bush (Figure 39).

The resulting RingRail would not be a pure ring; at Clapham Junction it would reverse (or, in effect, trains would run in both directions from here round the circle). This is necessary because Clapham Junction is such an important connecting point, which otherwise would be bypassed. It would link with the existing North London line between Willesden and Dalston Junctions, and this would in effect provide two outer spurs to Richmond and North Woolwich. But in addition, as long discussed, this line would be projected at its eastern end to provide a direct link with London City Airport and to pass under the river to

[30] Cervero 1989, 1991.

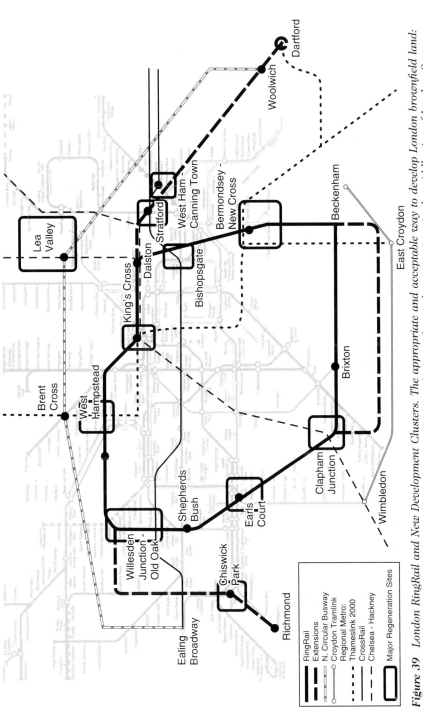

Figure 39 *London RingRail and New Development Clusters. The appropriate and acceptable way to develop London brownfield land: use of neglected and wasted rail and other land, around planned new rail interchange stations in the middle ring of London. Source:* London Railway Development Society and Transport 2000.

Woolwich Arsenal,[31] where trains could be projected onward to Dartford and Ebbsfleet – a critical investment for the Thames Gateway development. Finally, the existing electrified Gospel Oak to Barking line would constitute yet a further outer "wing" of the system.

The wider non-transport benefits of RingRail would be immense, since it would connect some of the most deprived areas of inner London – Hackney, Newham, Peckham, Brixton (with a new station) and Harlesden – giving them wider access to the rest of the London job market. In addition, the eastern projection could directly serve some of the most important development sites in Thames Gateway, including Stratford and Ebbsfleet stations on the Channel Tunnel Rail Link, the exhibition centre in the Royal Docks and Woolwich Arsenal. Through a new station north of King's Cross, it could even directly serve the new St Pancras terminus of the Channel Tunnel Link. Some problems would remain of interchange on its north side with the underground system – at Belsize Park, at Camden Town and north of King's Cross. There are solutions, but they would be relatively expensive. Perhaps the most attractive would be a link with the Northern Line, involving a people-mover, at Camden Town.

Though a relatively self-contained system, RingRail could be further enhanced on its south side by an outer loop via Sydenham, Crystal Palace, Streatham Hill and Balham to Clapham Junction. This would require only minimal investment at New Cross Gate. In any event, much of this outer loop is already served by the existing service from Beckenham Junction to Victoria, which will soon connect with the new Croydon Tramlink light rail system, due to open in 1999. And, in effect, Croydon Tramlink will form an outermost orbital system for south London, linking Beckenham via Croydon with Wimbledon – and thus with the southern-most section of the District Line north to Earl's Court, which here serves as an orbital route linking directly with RingRail at Earl's Court.

Thus, rather extraordinarily, London could acquire a system as good as Orbitale at bargain prices, and with minimal disruption. There would, however, remain a problem: south London would be splendidly served, by as many as three separate orbital loops; north London in comparison would have only one, running through the inner suburbs but leaving the outer suburbs, with their intensive orbital movements, without service. The irony here is that south London would have an excellent rail service and a poor road system, while north London, with its improved North Circular, would have the reverse.

But the North Circular is the potential solution. For it could provide the basis of an outer orbital public transport system. There are two possibilities. The first, which seems inherently less attractive both on cost and physical feasibility, would be to construct an elevated light rail, cantilevered above the existing central reservation on the North Circular. The other solution, which would be far simpler and equally effective, would be to complete the planned improvement of the North

---

[31] The twin underground lines under the Thames could be shared by Jubilee Line trains, using an existing stepplate junction at North Greenwich which has been provided as part of the Jubilee Line extension; this would give great benefits in accessibility to the Woolwich Arsenal regeneration site.

Circular between the A41 at Brent Cross and the A10 at Tottenham, but to couple this with creation of continuous HOV (High-Occupancy Vehicle) lanes, reserved for buses and taxis and shared cars; and to create on these lanes an electronic guideway system of the type now proposed for introduction in Liverpool, whereby electrically powered buses would be automatically guided, giving them the riding characteristics of very superior light rail vehicles at a small fraction of the capital cost. These electric buses could leave the North Circular to connect the major centres along the North Circular – Ealing Broadway, Neasden, Brent Cross, Wood Green, Ilford – which are important employment and service nodes as well as interchanges with radial public transport. A possible projection southward from Wood Green, using a long-abandoned rail right-of-way, could link with the Gospel Oak–Barking line at Seven Sisters.

In terms of timing, it should be possible to construct RingRail and its connections, plus of course the Croydon Tramlink, by 2005 and then go on to complete the North Circular Guideway by 2010. Thus London would have a system providing a high level of accessibility between almost all its major nodes. Further, new interchanges with the radial system would permit the creation of new mixed-use high-intensity activity centres where presently, because of the lack of connectivity, there is almost an urban wasteland; most notably at Chiswick Park and Old Oak in west London, on the King's Cross Railway lands and at Shoreditch–Bishopsgate. Similarly, on the North Circular Guideway, there are opportunities for intensification around the emerging Neasden Retail Park and in the Angel Road area of Enfield.

This is important, because a key element in a balanced portfolio approach to sustainable development would be to create new urban development opportunities within London in areas which presently, because of poor accessibility, are underdeveloped. An Orbitale system for London would create these opportunities at surprisingly low cost in comparison with larger rail projects planned at the present time, notably the proposed Regional Metro. But this, too, has a vital place in generating new patterns of accessibility as platforms for new urban development, both brownfield and greenfield. In London, the interchanges between the two systems would provide the keys to unlocking major regeneration projects.

NEW CONCEPTS IN PUBLIC TRANSPORT: (2) THE REGIONAL METRO

During the 1990s, a number of European cities have developed a radical extension of the concept represented by the Paris RER or the German S-Bahn systems: the Regional Metro. West of Stockholm, as proposed in the 1991 Mälardalen Regional Plan, new lines around Lake Mälar – Mälarbanan, Svealandsbanan and Grödingebanan – will provide a ring route connecting the capital with a series of medium-sized cities within a 100-km range, including Södertäjle and Eskilstuna on the south side, Enköping and Västerås on the north side, and Örebro at the west end of the lake. This appears to be the first case in which a regional development plan has been deliberately structured around the existence of high-speed links.

The new international Metro planned for the Öresund region, which will use the new link under construction and due to open in 2000, will link the Danish city of Roskilde to the west of Copenhagen, via Central Copenhagen and Kastrup international airport, with Malmö and Lund in southern Sweden. In northern France a regional system for the Nord-Pas de Calais region (TER) is under discussion, linking Calais and Boulogne with Lille and Roubaix. The Swiss national railways have developed a new system of fast trains linking all the major cities, with guaranteed connections.

LONDON'S REGIONAL METRO

London is developing its own concept of the Regional Metro,[32] which might better be called a Regional TGV (Figure 40). Like the other examples, but even more radical in concept, it essentially consists of very high-speed lines, with trains running at 125 miles per hour (200 km/h) or more, and connecting right across London to link cities and towns up to 80 miles (130 km) distant on either side.

The first, *Thameslink 2000*, is partially open but will be complete by 2003. Coming from the north of London it will have three branches, from Bedford on the Midland Main Line, Peterborough on the East Coast Main Line, and Cambridge and King's Lynn. These will converge at King's Cross into an underground link serving the west side of the City of London (Farringdon, City Thameslink and Blackfriars) before branching out again to serve destinations south of London: Dartford, Tonbridge/Ashford, Gatwick/Brighton and Eastbourne, and Sutton/Wimbledon.

The second line is odd, and does not even feature on the maps in London Transport's 1996 report. It is the *Channel Tunnel Rail Link* from St Pancras, which will provide for high-speed domestic trains running directly down the West Coast Main Line from Rugby, Northampton and Milton Keynes – upgraded at least to a 125 miles-per-hour (200 km/h) standard and more probably to 140 miles per hour (225 km/h) – via the St Pancras hub, to Ebbsfleet and Ashford. Essentially this will have two branches: one, diverging from the high-speed link at Ebbsfleet, serving the Medway towns of Rochester, Chatham and Gillingham and then the North Kent coastal towns of Whitstable, Herne Bay, Margate and Ramsgate; the other, diverging at Ashford, which will have three branches, one to Hastings, one to Canterbury and Ramsgate, a third to Folkestone, Dover, Deal and Ramsgate. In fact, these last two branches connect with the North Kent line to form a continuous rail loop around the north-east corner of Kent.

Both Thameslink 2000 and the West Coast Main Line/Channel Tunnel Rail Link will be ready by 2007, since the critical station works for both lines at St Pancras have to be built simultaneously. The other schemes are longer-term. *CrossRail*, an ambitious scheme for a deep-level full-size underground line connecting Paddington, Bond Street, Farringdon and Liverpool Street–Moorgate, would link together commuter lines coming from the west of London, one branch from Reading and Slough, the other from Aylesbury and Amersham, to the Great

---

[32] London Transport 1995, 1996.

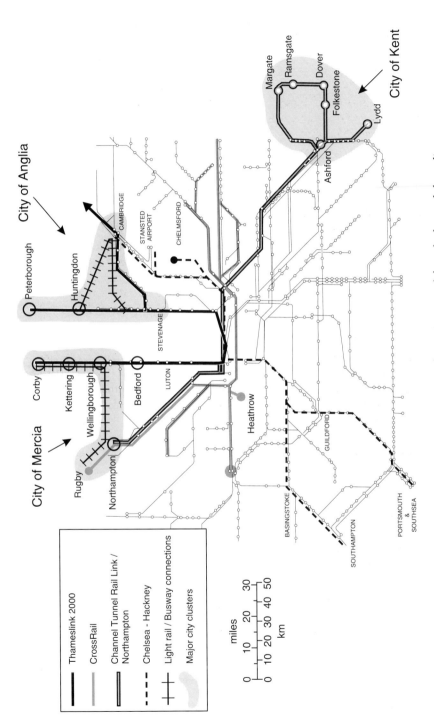

Figure 40 *Regional Metro and New City Clusters. London Transport's 1995 proposal for a high-speed, long-distance system interconnecting under central London; Thameslink 2000, CrossRail and Chelsea–Hackney would be the three lines, but the Channel Tunnel Rail Link would effectively provide a fourth. Source: London Transport.*

Eastern Line out to Shenfield. London Transport are now suggesting that it could also incorporate trains on the West Coast Main Line from Rugby, and the Heathrow Express, on the west side, and the two Southend lines on the east side. Because of its expense, CrossRail has been postponed until after the completion of Thameslink 2000, but it seems a high priority for completion between 2005 and 2010.

The longest-term scheme of all is the *Chelsea–Hackney* link. Originally planned as a conventional underground line, it is now being redesigned to bring trains from the South West Main Line (Southampton, Basingstoke) on to a diagonal line connecting Chelsea, Victoria, Piccadilly Circus, Tottenham Court Road and King's Cross, then exiting through north-east London to serve Chingford and Epping. There is no date for its construction: it is unlikely to be before 2010–2020.

As London Transport argue, the Regional Metro could have dramatic effects on the development of the entire London region, at least as momentous as those wrought by the extensions of the underground system in the 1920s and 1930s. The spatial impact would however be quite different: by offering very fast access to selected points, it would extend the catchment area of central London, both for commuters and shoppers/leisure travellers, to a series of destinations within about 80 miles (130 km). But this extension would not be at all even: it would be highly punctiform, serving selected stations. It would thus provide the potential basis for a similarly punctiform settlement strategy, based on pyramids of density around the stations. Thus it would facilitate a process that has been continuing apace since 1950: the translation of the South East of England into a polycentric metropolis based on what the Dutch call the principle of concentrated deconcentration.

POLICY RESPONSES: LAND USE PLANNING

That reminds us that there is another approach – not as an alternative to transport policies, but as a complement to them – through land use. We can seek innovative solutions from three sources. First, as said, we can turn back to the wisdom of the past, and ask ourselves how the great planning figures of history managed similar challenges. But, as part of that exercise, we need also to ask what things they got wrong, including especially the things that worked then but do not work so well now, and may therefore need re-analysing and re-interpreting. Second, we can go to the academic research, and ask whether it is currently discovering anything useful to us. And third, we can ask whether there are any good contemporary working examples around. When we put these three approaches together, we not only get an answer, we get a very clear answer.

THE WISDOM OF THE PAST

We can and we should start by going back to Howard. For, while the world has changed out of all recognition, his message still has a startling, almost surreal relevance to us in the Britain of 1998.

As seen in Chapter 7, the Three Magnets Diagram proves as relevant in 1998 as it did in 1898. But the overwhelming difference is that people have discovered for themselves just how attractive the town–country magnet is. They have flocked out to it in millions. The problem is that the planning system failed to provide for it. People did not hark back to Howard's Social City. Instead, they – and still more their local political leaders – stuck their heads in the sand, saying they did not want the growth. They failed to stop it, of course, but because they failed to provide for it as Howard suggested, they have been stuck with the worst of all worlds.

We can also look back to Chapter 6, at the Scandinavian approach. Both in Copenhagen and in Stockholm this was to string new settlements like beads along public transport routes, in the case of Copenhagen a tram system and then a suburban train system, in Stockholm a new underground railway.

The morals we can draw from history are these. First, grouping people and jobs close together in reasonably self-contained units, as Howard proposed and the British tried to do in their new towns after World War II, is a reasonable objective – particularly if, at a micro-scale, homes and jobs are intermixed. Howard's intuitively based size for his Garden Cities – 32 000 – was not bad in terms of transport sustainability, despite the fact that it was produced at a time when the motor car was a novelty, but it proved difficult to combine this size of city with an adequate range of jobs and services. But, of course, everyone had forgotten the original Social City diagram. Second, it was also right to try to keep the new British towns outside the London commuting orbit, even if some commuted and more might later do so; you could keep the new towns quite self-contained on a day-to-day basis (and the evidence is that some at least of the commuters found local jobs after a while), while providing good access to London for less frequent business contacts, which have become more and more significant with the growth of service industry jobs and the decentralisation of back offices into the new towns. Finally, putting homes and jobs in patterns of high linear density along strong public transport spines, as the Swedes did in their Stockholm satellites of the 1950s and 1960s, is also right – especially if, again, some jobs were provided close to homes, as the Swedes tried to do, and if wide – in fact increasingly wide – green wedges are thus created between these corridors of urbanisation. All represent partial answers, by no means mutually exclusive, but hardly any place seems to have combined them in a package.

We should also remember that the Mark One London new towns, and the Scandinavian satellites, were devised for a much poorer society, in which the majority of people would be renters of social housing provided by large monolithic agencies; they had no choice. Nor did they have much choice in transport, for car ownership levels were extremely low and expected to remain so. There is clear evidence that all the calculations of the Stockholm planners were upset by the rapid rise of car ownership in the 1960s. But this is coupled with the fact that Stockholm is a relatively small and compact city of only about a million and a half people. Even if people could be persuaded to continue to use the Tunnelbana for the radial journeys to the city centre, it made little sense for orbital journeys when they could get into their comfortable Volvos and use the ring road the planners had provided to get freight around the city. This suggests that the structure has to be more encouraging to public transport use, perhaps by eliminating as far as possible the need or desire for non-radial journeys.

THE CONTEMPORARY ACADEMIC WISDOM

The second source of wisdom, or inspiration, is the current academic work. The starting point here must be the well-known study by Newman and Kenworthy,[33] which makes it clear that, overall, European cities are denser than either Australian or American ones, and that this is systematically associated with a higher usage of public transport and with lower energy consumption per capita: average petrol consumption in American cities is nearly twice as high as in Australian cities and four times higher than in European cities. Differences in petrol prices, income and vehicle efficiency explain only about half of these variations. What is equally significant, however, is the urban structure: cities with strong concentrations of central jobs, and accordingly a better-developed public transport system, had much lower energy use than cities where the jobs were scattered. Overall, Newman and Kenworthy found a strong relationship between energy use and the use of public transport, especially rail, and provision for the car. In European cities, 25 per cent of all passenger travel is by public transport and only 44 per cent use a car for the journey to work. The importance of walking or biking in these more compact cities is highlighted by the fact that 21 per cent use these modes for their work trip. In Amsterdam the proportion rises to 28 per cent and in Copenhagen to 32 per cent.

Newman and Kenworthy's conclusions have been seized upon in some important policy prescriptions. Thus the well-known European Commission Green Paper[34] assumes an ideal urban form, represented by the traditional compact European city which depends on short distances to work and to shop and is supported by generous investment in public transport, which uses fewer non-renewable resources and which creates less pollution than the scattered Anglo–American–Australian urban form. However, Michael Breheny has questioned the Green Paper for what he regards as an obsession with density.[35] His work, and that of others, suggests that moderate densities may be perfectly satisfactory; it is not clear that it is necessary to pack everyone into existing cities.

Their work has been criticised methodologically[36] and ideologically.[37] Gordon and Richardson argue that their analysis is faulty, that they wrongly diagnose the problems, and that their policy and planning prescriptions are inappropriate and infeasible. In American cities, Gordon and Richardson and their co-authors argue, suburbanisation of employment together with population has actually reduced, not

---

[33] Newman and Kenworthy 1989a, 1989b, 1992.
[34] Commission of the European Communities 1990.
[35] Breheny 1992.
[36] Schipper and Meyers 1992.
[37] Gordon and Richardson 1989; Gordon et al. 1991.

lengthened, commuting times and distances: people have stopped making long suburb-to-city trips and are making short suburb-to-suburb trips instead. Brotchie et al.[38] reach the same conclusion for Australian cities. Newman and Kenworthy, in new research, defend themselves on these scores, but there are some real remaining doubts about their methodology.[39]

In any case, we can doubt – for reasons already spelt out – whether it will be sufficient to argue simply in terms of urban consolidation. We need to look at ways of achieving sustainable urban forms for new development on greenfield land. British geographers, interestingly, have established some kind of international lead in this topic: researchers like Susan Owens at Cambridge, Michael Breheny at Reading, and David Banister at University College London together make up a very formidable group.[40] And all this work seems to tell a very consistent story.

Owens suggests that a sustainable urban form would have the following features. First, at a regional scale, it would contain many relatively small settlements, but some of these would cluster, to form larger settlements of 200 000 and more people. Second, at a sub-regional scale, it would feature compact settlements, probably linear or rectangular in form, and with employment and commercial opportunities dispersed to give a "heterogenous", i.e. mixed, land use pattern. Third, at the local scale, it would consist of sub-units developed at pedestrian/bicycle scale; at a medium to high residential density, possibly with high linear density, and with local employment, commercial and service opportunities clustered to permit multi-purpose trips. Her work strongly suggests that a cluster of small settlements may be more energy efficient than one large one; the optimum upper limit would be 150 000–250 000; that linear or at least rectangular forms will be the most efficient; and that though densities should be moderately high, they need not be very high to be energy-efficient. Thus, a density of 25 dwellings per hectare (10 per acre), which in terms of future composition might translate into about 40 people per hectare or 16 per acre, would allow facilities with a catchment area of 8000 people to be within 600 metres of all homes, and a pedestrian-scale cluster of 20 000–30 000 people would provide a sufficient threshold for many facilities without resort to high densities, which actually might be energy inefficient. These arrangements accord with what Stockholm achieved around the Tunnelbana stations in the 1950s, though at UK new town densities. Owens also points out that district heating systems are viable at moderate densities of 30/37 dwellings per hectare (12–15 per acre); on a greenfield site this break-even density would be even lower, particularly if land uses are made heterogeneous.

In his work with Ralph Rookwood[41] Breheny gives some theoretical illustrations of how sustainability would be developed at different scales and in different

---

[38] Brotchie et al. 1995, 382–401.
[39] Kenworthy et al. 1997; Newman and Kenworthy 1997.
[40] Banister 1992, 1993; Banister and Banister 1995; Banister and Button 1993; Breheny 1991, 1992, 1993, 1995a, 1995b, 1995c; Breheny and Rookwood 1993; Breheny et al. 1993; Owens 1984, 1986, 1990, 1992a, 1992b; Owens and Cope 1992; Rickaby 1987, 1991; Rickaby et al. 1992.
[41] Breheny and Rookwood 1993.

geographical contexts. All of them feature settlements of different sizes, strung like beads on a string along public transport corridors which range from bus routes up to heavy rail systems (Figure 41). Again, there is a very strong similarity to what Danish and Swedish planners were attempting in the 1950s and 1960s.

More recent work, co-ordinated by David Banister and financed by the Economic and Social Research Council, is more empirical. Peter Headicar and Carey Curtis at Oxford Brookes University looked at a crucial question: whether, in terms of car dependence, it matters where you put new suburban development. They conclude that it very much does. Most people moving to new suburbs close to big towns and on corridors with good public transport, like Botley and Kidlington outside Oxford, commuted by car before the move and even more did so afterwards. However, the proportions commuting by car, both before and after the move, are lower than for people who moved into smaller country towns like Bicester and Witney, further away from Oxford. And at Didcot, located on a good rail line, higher proportions used the train – though even there, four in five commuted by car. In terms of total car usage, expressed as miles per week, the near-in suburbs of Kidlington and Botley generated the least travel, while the more distant towns of Bicester and Witney generated two to three times as much. Of the three country towns Didcot generated the least car travel, close to the Botley total, and the biggest amount of rail travel. These differences, the researchers concluded, are linked not just with size and density, but with location within a broader sub-region.[42] The policy implication seems to be this: it is best to locate developments on strong public transport corridors, close to medium-sized towns that will continue to provide the main sources of employment, and rail corridors offer particularly good prospects.

THE WISDOM OF CONTEMPORARY PRACTICE

Thirdly, we have contemporary practice. From America we have some interesting ideas from Peter Calthorpe, a Californian English *émigré* architect–planner.[43] He proposes walking-scale suburban developments around public transport stops, clustering some job and service opportunities at the nodes, and with high-density single-family housing built in traditional terraces with street parking (Figure 42). His concept, which he calls Transit-Oriented Development (TOD), bears an astonishing physical resemblance to the ideas of Breheny and Rookwood, developed independently and published in the same year. Californians seem to like it; he has developed whole neighbourhoods in San Jose, the capital city of Silicon Valley, and his ideas have now been made a mandatory part of the General Plan for the state capital of Sacramento.

Recently, the Netherlands have taken a worldwide lead in trying to integrate land use and transport planning, within an environmental strategy, at a national level. The *Fourth Report (EXTRA) on Physical Planning in the Netherlands*

---

42   Headicar and Curtis 1996.
43   Calthorpe 1993; Kelbaugh et al. 1989.

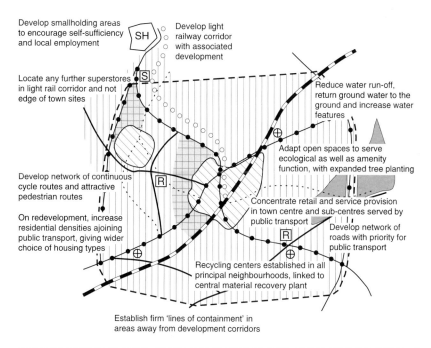

Develop smallholding areas to encourage self-sufficiency and local employment

Develop light railway corridor with associated development

Locate any further superstores in light rail corridor and not edge of town sites

Reduce water run-off, return ground water to the ground and increase water features

Adapt open spaces to serve ecological as well as amenity function, with expanded tree planting

Develop network of continuous cycle routes and attractive pedestrian routes

On redevelopment, increase residential densities ajoining public transport, giving wider choice of housing types

Concentrate retail and service provision in town centre and sub-centres served by public transport

Develop network of roads with priority for public transport

Recycling centers established in all principal neighbourhoods, linked to central material recovery plant

Establish firm 'lines of containment' in areas away from development corridors

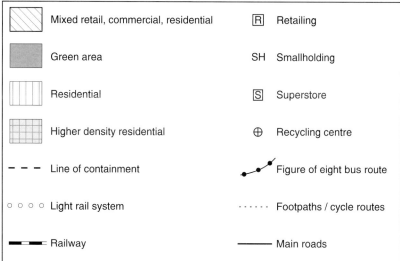

Mixed retail, commercial, residential		R	Retailing
Green area		SH	Smallholding
Residential		S	Superstore
Higher density residential		⊕	Recycling centre
– – – Line of containment			Figure of eight bus route
○ ○ ○ ○ Light rail system		· · · · ·	Footpaths / cycle routes
▬▬▬ Railway		———	Main roads

Figure 41 *Breheny and Rookwood's Sustainable Development. An example from their 1993 paper, showing the application to a mixed urban–rural area: clusters of mixed-use developments, each of limited size, along public transport spines; the intervening countryside preserved.* Source: Breheny and Rookwood 1993.

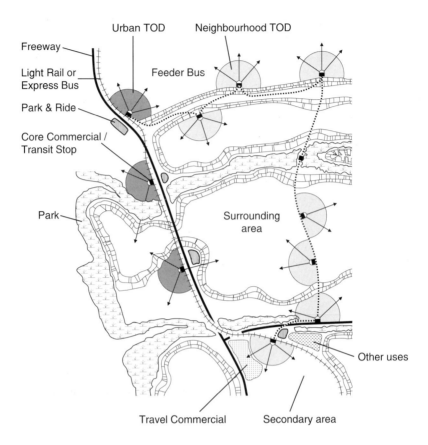

Urban TOD   Neighbourhood TOD

Freeway

Light Rail or Express Bus

Feeder Bus

Park & Ride

Core Commercial / Transit Stop

Park

Surrounding area

Other uses

Travel Commercial   Secondary area

Figure 42   *Calthorpe's Transit-Oriented Development. Also published in 1993, this diagram from a California architect–planner was conceived quite independently to Breheny and Rookwood's work – yet there is an astonishing similarity. The experts, it seems, are agreed.* Source: Calthorpe 1993.

identifies a policy that aims to cope with growth pressures and to improve the quality of urban life and reduce car traffic in cities and urban regions, through an integrated approach encompassing traffic and transport policy, environmental policy and physical planning policy. The key is to concentrate residences, work areas and amenities so as to produce the shortest possible trip distances, most being possible by bicycle and public transport. So housing sites are being sought first in the inner cities, next on the urban periphery and only in the third place at more distant locations; wherever the sites are found, availability of public transport will be a key factor. Businesses and amenities are planned by relating their user requirements to location features. Those activities involving a large number of workers or visitors per hectare, such as offices oriented to the general public, theatres and museums, are rated A-profile, that is they should be located close to city-centre stations. B-locations are those with both good station access and good access to motorways, making them suitable for access by both car and public transport; activities suitable for location here include hospitals, research and development, and white collar industry. C-locations, close to motorways, are suitable only for activities with relatively few workers and visitors per hectare and with a need for high accessibility by car or truck. Associated with this, the Report calls for integrated transport/land use planning so as to enhance the role of public transport, including restrictions on long-term parking places, associated with the provision of good public transport.[44]

The Dutch approach is stimulating a great deal of interest and even imitation elsewhere in Europe.[45] But it might make equal or more sense to relieve pressure on Randstad Holland – the horseshoe-shaped ring comprising Rotterdam, The Hague, Leiden, Haarlem, Amsterdam and Utrecht – by promoting moderately-sized, moderate-density cities elsewhere in the Netherlands: a policy of the 1960s, later abandoned.

There is an equally remarkable initiative coming out of the UK. The development framework for Thames Gateway, east of London, represents a new scale in thinking: a discontinuous development corridor more than 40 miles (64 km) long, based on the new high-speed train link from London to the Channel Tunnel, which will have concentrations of employment around two planned stations, and with dense local rail travel along a new "Thames Gateway Metro" in between.[46] In many ways, it represents the ideas of Breheny and Rookwood, or Calthorpe, realised in a real-life context. We return to it in Chapter 9.

These building blocks give us all we need to develop sustainable packages of land use and transport policies. Further, they will actually be happening in different parts of the world, so that we will be able to see how well they work, and thus learn from each other's successes and mistakes. We shall be able to see how Calthorpe's notions of sustainable urbanism, the Dutch ABC policies, and the British corridor approaches and regional metro are actually working – and we shall be able to study the development outcomes of the London regional metro with the very similar system now under construction around Stockholm. But here in Britain we do not have to wait for the others; we can develop our home-grown solution. This we will try to spell out in more detail in Chapter 9.

---

[44] Netherlands 1991.
[45] e.g. London Planning Advisory Committee 1993.
[46] GB Thames Gateway 1995.

SUSTAINABLE SOCIAL CITIES OF TOMORROW

KEY STRATEGIC POLICY ELEMENTS

We know, then, the strategic policy elements on which we can build sustainable Social Cities of the kind that Howard would have recognised, but that answer to the needs of the twenty-first century. These building blocks are 12 in number:

1. *Develop Urban Nodes*: Within London and other large cities, systematic efforts should be made to create new accessibility nodes by selective investment in new transport links. This would create a more polycentric model which would balance flows on public transport corridors. Since, as a rule, radial transport is much better developed, and more convenient, than orbital transport, the major efforts should be made in improving these latter links and then developing nodes of concentrated higher-density development at their intersections with the existing radials, which often are surprisingly weakly developed. Here, the Swedish principle of pyramids of density would be employed: shopping, services and entertainment should be strongly concentrated around the interchange stations, with higher-density apartment living above and around these nodes, and with single-family homes served by longer walks or bus links outside a radius of approximately half-a-mile (1 km).

2. *Selective Urban Densification*: Urban compaction, or intensification, is acceptable, in fact desirable, in order to help urban economic regeneration, make cities more lively, induce less use of the car and protect open countryside.

3. *No Town Cramming*: However, densification must be compatible with good urban quality. This will certainly mean the rigorous protection of urban green spaces. It may mean restrictions on the scale of higher residential densities; otherwise the policies will be completely counter-productive, causing even more people to leave. This is a particular problem around many suburban public transport nodes, which were first developed a century or more ago, and from the start developed a structure based on single-family villas rather than high-density apartment blocks. London has many such places – Ealing Broadway, Wimbledon, Harrow on the Hill, Croydon, Bromley South – and most of them still offer a high-quality residential ambience. Their residents

would fiercely oppose any attempt at large-scale redevelopment and densification, and even conversion of larger houses into flats may evoke NIMBY opposition. In general, therefore, it may be easier and better to develop "new towns intown" at new interchanges which presently are under-developed.

4. *Strategic Provision for Greenfield Development:* Because of these limits, selective densification can never presume to provide for more than a half, at the very maximum, of the national housing demand. It may be more than half in some regions with loose urban structures, such as the North West; it will be a good deal less in other regions, such as the South West, where there is little available land within the urban envelopes and little likelihood of much more coming on stream. As a result, we should be providing as a contingency for about 55–60 per cent of all development to be taking place in the form of new greenfield development. The scale of greenfield development will be so pervasive and widespread that purely local solutions will not be effective or adequate; the strategies require working out at a regional and then at a sub-regional level, transcending the boundaries of local planning authorities.

5. *Distance:* Developments must be far enough from the largest metropolitan cities to guarantee, so far as this is ever possible, that they will be reasonably independent of it in terms of everyday movements. In Howard's day, a distance of 35 miles (56 km), separating Letchworth from London, was more than enough. Now Letchworth is comfortably within the range of West Anglia and Great Northern Railway's electric suburban trains. Five thousand and more people commute daily from Peterborough, one of the Mark Two new towns designated in the late 1960s, because high-speed trains bring it within a mere 40-minute journey of King's Cross. Within a few years, as Inter-City 125s give way to Inter-City 225s and 250s, travelling at 140 or perhaps 150 miles an hour (225–250 km/h), much of England will become potentially a single vast suburb, as H.G. Wells predicted just three years after Howard wrote.[1]

   But that does not mean we have to abandon the task. While some will choose the freedom that the new trains will give them, the majority will settle for working more locally. The further we place the new developments from London or Birmingham or Manchester, the greater will be this degree of self-containment. As a rule of thumb, the minimum distance for intensive large-scale development should be roughly the location of the Mark Three new towns, that is 50–90 miles (80–140 km) from London. In effect, this means a continuation and extension of the approach of the authors of the 1970 *Strategic Plan for the South East,* when they modified the corridor strategy in the Planning Council's 1967 *Strategy:* the corridors should have limited and selective development within the 50-mile (80-km) ring, with larger-scale and more intensive development beyond that.[2]

6. *Top-Quality Linkages:* However, because some will commute (and indeed some must commute; dual careers, and greater specialisation, mean that the

---

[1] Wells 1901.
[2] GB South East Economic Planning Council 1967; GB South East Joint Planning Team 1970.

world of work is no longer as simple as in Howard's day), then it will be right to ensure that they can make their journeys as quickly, as easily, and above all with least resource cost and pollution, as possible. That clearly means relying on the high-speed train network, now in course of improvement and upgrading, and on the new concept of the London Regional Metro, which can be combined to produce a network of services serving not merely a London terminus, but a variety of stops within London's business heart. The first three of these services, within the first few years of the new century, will be the upgraded West Coast Main Line, connected directly at St Pancras to the Channel Tunnel Rail Link to East Kent; and the Thameslink 2000 service connecting Bedford and Peterborough and Cambridge/King's Lynn with Gatwick and Brighton and Eastbourne. These services in effect extend the potential London commuter ring somewhat further than was feasible in 1970, with an emphasis on places between 60 and 85 miles (100–140 km) from central London.

7. *Clustered Development*: Nothing is more clear, in the recent literature that we summarised in Chapter 8, than the remarkable degree of agreement among researchers and practising planners, both in the UK (Breheny, Rookwood, Owens) and in the USA (Calthorpe), on the most sustainable urban form. Basically, it is a linear version of Howard's Social City, with relatively small walking-scale communities (population: 20 000–30 000) clustered along public transport routes, especially rail, light rail or guideway. Such communities might be rather more thinly clustered in locations closer to major metropolitan centres (e.g. roughly within 50 miles or 80 km) than locations further-distant, where they could cluster more closely to form what would in effect be regional agglomerations: the concept developed in the 1970 *Strategic Plan for the South East*, but there applied to a variety of locations at different distances from London.

8. *Town Expansions and New Towns*: Within these corridors, the clusters would contain a mixture of different kinds of development. Medium-sized and smaller towns along such corridors can be expanded (with localised, acceptable compaction) around good quality public transport nodes, typically providing interchanges between longer distance rail or light rail stations and local distributor systems. But new towns may also be an appropriate solution; indeed, they may provide the only alternative to a "pepperpotting" of development which will make no strategic sense and will be maximally unpopular at local level.

Such new communities, however, could be and probably will be very different from either the Mark One new towns of 1945–50 or the Mark Three new towns of 1961–70. As already suggested, they are likely to consist of small semi-self-contained, physically separate, mixed-use units of 20 000–30 000 people, akin to Ebenezer Howard's original formulation of Garden Cities, but clustered – as again he proposed – into larger units of up to 200 000 or 250 000 people along the transport corridors, which will also contain expansions of existing towns. This is an issue that will demand careful regional (and possibly inter-regional) consideration.

9. *Density Pyramids*: Around these stations, as within the major urban areas, we should follow the principle the Swedes used in the Stockholm satellites of the 1950s: higher densities, including a preponderance of apartment blocks, within walking distance of the stations; and shopping and other services concentrated immediately next to the stations, making it possible to make a transition train–shop–home. This principle should apply both to the existing towns that are being expanded, and to the new towns.

10. *Variation According to Geography*: As Howard reminded us on his famous diagrams, all theoretical principles and plans must be modified by the facts of the land. Though in general development should be clustered along transport corridors, that principle should not apply in Areas of Outstanding Natural Beauty; no one would contemplate large-scale development in the New Forest or the Cotswolds or the Weald of Kent, for instance. Conversely, a town with physical room to expand should probably be encouraged to do so, even if it is not directly on a rail line.

11. *Areas of Tranquillity*: The Council for the Protection of Rural England have drawn attention to the erosion, over the past 40 years, of areas of deep countryside which are reasonably tranquil in terms of low noise levels. We agree that it should be a prime aim of policy to protect our remaining tranquil areas to the maximum degree possible. To that end, we need to go back to the principles underlying the 1970 *Strategic Plan for the South East*: each region should be broadly divided into zones for priority regeneration and development, normally occupying only a small part of each region, and much larger areas of protected tranquil countryside. This is one of those examples, not infrequently found, when a good policy may also make good politics.

    Within these more rural areas, if decentralisation of jobs and people continues to place pressures on the local housing market, it may be appropriate to focus growth into key villages. This may have the merit of avoiding unpopular multiple village expansions, as well as supporting public transport and services. However, in zones of tranquillity the general principle should be that development is restricted to meeting local needs, and this demands that special attention is given to prioritising provision of social housing.

12. *Stimulate Remote Rural Areas*: Outside the zones of outward pressure from the cites and conurbations, rural decentralisation should actually be encouraged. Such decentralisation has alone sustained such villages in recent years, helping them maintain their role as service centres to the indigenous local people, especially the older and lower-income ones. But such development must be absorbed sensitively through village plans.

THAMES GATEWAY: SUSTAINABLE DEVELOPMENT PROTOTYPE

In Thames Gateway, east of London, British planners have at last invoked the spirit of the great American planner Daniel Burnham: Make No Small Plans. The development framework for the area, published in 1995, represents a new scale in thinking: a discontinuous development corridor more than 40 miles (64 km) long, based on the new high-speed train link from London to the Channel Tunnel, which will have concentrations of employment around two planned stations, and with dense local rail travel in between.[3] Not merely is it huge in scale, larger than any similar earlier project (Figure 43), but it also is postulated on a massive gamble as to the impact of a new transport technology.

Just as in the past transport has proved to be maker and breaker of cities, in Colin Clark's telling phrase,[4] so here. The spatial impact will be quite different from traditional urban rail systems like the London Underground or Paris Metro or Stockholm Tunnelbana: it will be to telescope travel times to places in the critical range 60 to 80 miles (100 to 130 km) from London. But Clark's postulate has an even profounder significance here. The establishment of London's airport at Heathrow, the result of a hasty wartime decision in 1943, has profoundly influenced the postwar growth of the metropolis and its region, through its magnetic effect on business, in the entire corridor that stretches from the West End out to Reading and beyond – perhaps even to Bristol. We now have a major technological discontinuity in the form of the high-speed train: in effect a new transport technology, its effects are likely to be equally momentous. It comes into its own in densely urbanised regions with a large number of major conurbations within about 300 miles (500 km) of each other, since that is the distance over which high-speed trains prove to be highly competitive against air travel. The north-west corner of Europe – the area the European Commission calls the Centre Capitals Region, embracing London, Paris, Brussels, Amsterdam and Cologne–Bonn – is just such a region, and within it the train will rapidly become the normal means of inter-city travel for business and leisure purposes.

Further, its urban effects will be profound: the trains will re-centralise business in the hearts of the cities, though not necessarily right at the train stations, but they may also encourage selective large-scale developments on the fringes of major metropolitan areas. This has already occurred at Shin-Yokohama, on the original Tokaido Shinkansen outside Tokyo, and Massy in the south-western part of the Région Ile-de-France, at the critical junction between the Atlantique and the new Interconnection, where a major development is planned as the gateway to the so-called Cité Scientifique Ile-de-France Sud, the French equivalent of the M4 Corridor. *The Thames Gateway Planning Framework*, published by the Department of the Environment in 1995, exploits this by providing for two activity centres round the planned stations at Stratford and at Ebbsfleet, close to the M25 Dartford Crossing.[5]

These two locations, it can confidently be said, provide the two major anchors for employment and activity in the entire development, and their spacing, approximately 20 miles (32 km) apart, is just about right. One of the most important and most intriguing tasks for those who will plan the development of the corridor is exactly what mix of activities will go to Ebbsfleet, what mix at Stratford

---

[3] GB Thames Gateway 1995.
[4] Clark 1957.
[5] GB Thames Gateway 1995.

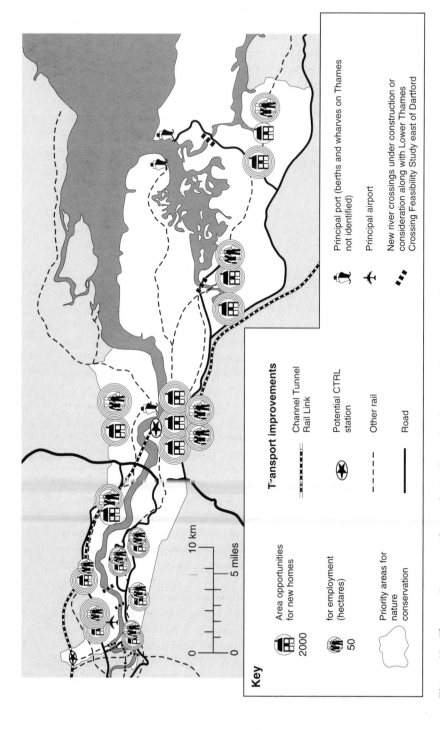

Figure 43 *Thames Gateway. Clustered mixed-use development along the new Channel Tunnel Rail Link, with main activity centres around two of the Link's stations, Stratford and Ebbsfleet.* Source: GB Thames Gateway Task Force 1995.

Key

Area opportunities
for new homes
⊞ 2000

for employment
(hectares)
50

Priority areas for
nature conservation

Transport improvements

Channel Tunnel
Rail Link

✈ Potential CTRL
station

Other rail

Road

Principal port (berths and wharves on Thames
not identified)

✈ Principal airport

New river crossings under construction or
consideration along with Lower Thames
Crossing Feasibility Study east of Dartford

0   5 miles
0   10 km

and what next door to Stratford in the Royal Docks. Already emerging are scientific research and higher education, entertainment and cultural activities, and exhibition and conference space: centres for human interaction on a vast scale.

These two centres or anchors will be connected by a strongly linear transport infrastructure, comprising the high-speed line, the domestic services that will also operate on that line, the existing rail lines north and south of the river – both of which are scheduled for extensive upgrading, and which could become elements of an RER-type system for London – and the main highways, the A13 north of the river, and the A206 and the A2/M2 south of it, which also are being comprehensively upgraded. And, tying these together, there will be cross-links: the two, later probably three, Blackwall Tunnels, an East London River Crossing, the M25 and the future Lower Thames Crossing, plus very probably a future cross-river rail link at or near Woolwich.

There are several remarkable features of this infrastructure. One is that most of it should be in place soon after the turn of the century. The other is that the investment package is a mixture of road and rail (and possibly also water; we should not neglect the possibility of express riverbus service), in which the non-road modes will have a very significant share. We have to be realistic about such matters, but in so far as it is possible to devise an environmentally friendly transport package as the basis for an environmentally friendly urban development, this surely is it.

There will of course be other activity centres. Some are already in existence, such as Chatham Maritime; others are potential, such as the Rainham Marshes. Between them will be extensive areas that are predominantly residential; not exclusively, because they will be mixed with manufacturing and warehousing and retail and open space and a great variety of other activities. Thames Gateway is thus a unique chance to design a model sustainable urban development on a huge scale, a model for the entire world. This is first because of the rail infrastructure, and second because – despite the huge magnet of central London next door – the existing settlement structure already provides a great deal of self-containment: about two-thirds of the existing residents already work locally, and it should be possible to increase that percentage, both by building up the strong activity magnets at the two ends, and also by stressing the smaller-scale local employment opportunities between them. Further, Thames Gateway is, so to speak, insulated from the desire for non-radial trips: to the north the Thames provides an effective barrier, with no fixed crossing downstream of the M25 bridge–tunnel at Dartford; to the south, the Garden of England will remain a low-density agricultural area with few opportunities of any kind. This is essentially the thinking of Owens, Breheny and others, actually realised in a development plan. Within 15–20 years we should be able to see how well it performs in sustainability terms.

Thames Gateway, as has been stressed, is not and will not be a Linear City, much as that phrase makes good journalistic copy. It will be linear: the Thames ensures that. It will be urban, though it will also be rural; there will be a complex intercalation of urban areas and intervening green belts, green strips, green wedges. But it will not at all be a single monolithic city; it will be something altogether more complex than that, a corridor with multiple centres large and small, with residential areas strung out between them. It will also have a great deal of intervening open space: the urban elements are separated by vast spaces of marsh or quarried chalk, always backed by the equally vast expanse of the tidal Thames. It is a very horizontal landscape, a very eastern English landscape: often savagely degraded, often very monotonous, but possessing a kind of tremendous drama or glory of its own, because of the huge mass of water and the feeling that this is indeed England's great corridor to Europe and the world.

The townscapes are equally important. They often give an extraordinary sense of history. There is the great street market at the branching of the ways of the great eastern post road at Stratford, the old medieval herring port next to the great space of the old ruined Abbey in Barking, the high street of Gravesend climbing so dramatically away from the river; all these and others are going to have to be conserved and enhanced. But they do not give us much clue as to what to do with the intervening space. This is going to be a job of designing not one new city, but a complex pattern of old towns, new towns, new suburbs; in fact, a great variety of different urban forms.

One can almost say, then, that the main elements of the design fall logically into place of their own accord. There will be parallel road and rail routes on each side of the Thames itself. The new rail link, and the associated commuter services, will provide an additional diagonal path linking these others. All these will be joined by many short cross-links, two of which will be longer and of regional importance: the M25 and the new East London River Crossing. There will be two strong activity anchors at the western end, one of which will form a kind of articulation or jointing structure with the Docklands development. And there will be one or more such anchors at the eastern end. Between them, and around them, areas of special character will provide constraints and opportunities which the urban design will have to recognise.

But what of the intervening spaces? A major objective of the entire exercise, after all, must be to provide homes to meet a part of the entire projected demand in the South East of England over the next 15 years: about 100 000 new homes, the equivalent of several medium-sized new communities, 30 000 of them in London, the remainder mainly in Kent. We are not, however, talking about either Mark One or Mark Two new towns; nor about the smaller, self-contained, privately financed new communities, like Stone Bassett and Foxley Wood, that became so fashionable with the housebuilders (but so much less popular with Secretaries of State for the Environment) during the 1980s. This is because geographical conditions in the corridor very powerfully condition what kinds of communities can be built: the available sites – at Barking Reach (Figure 44), Havering Riverside, and Thamesside Kent (Figure 45) – are each quite discrete, and will support developments typically of 5000–10 000 homes apiece.

What is really in question here is garden suburbs. By definition they will be suburbs, for three main reasons. First, however many jobs we can and should provide in the corridor itself, many of the people who come to live here will find employment in central London, while relatively few are likely to find suitable employment right on their own front doorsteps. Second, the linear form means that the predominant direction of movement will be east to west, or vice versa,

Figure 44 *Barking Reach. Site of the first major residential development in Thames Gateway: 6000 homes and associated jobs, now under construction between the Tilbury Line and the river.* Source: Peter Hall photograph.

Figure 45 *Eastern Quarry. The location of a major residential development in Thamesside Kent, close to the Ebbsfleet station on the Channel Tunnel Rail Link and its associated business park.* Source: Peter Hall photograph

along the spines. And third, they will be built by private builders working to sell their houses in the market, and the market shows that the majority of people are going to be looking for fairly conventional single-family housing with private garden space.

Historic Victorian garden suburbs work so well because they are deliberately and self-consciously railway suburbs, focusing on station and shops, but also so related to green space that they have a sense of arcadian, semi-rural calm. The Victorian commuter left his rural retreat to focus on the station that would take him into the bustling city; he returned to face back into the arcadian retreat. But clearly, latter-day garden suburbs are going to have to take account of a whole range of considerations that Victorian urban designers never had to worry about. The new concern for personal safety, which runs counter to every good principle of designing quiet pedestrian access to places, is just one. Another, very evidently, is that however hard you try to design suburbs to persuade people to use convenient public transport, large numbers of them are still going to commute by car. This means that what Kevin Lynch calls the path system will have to provide – both in functional and visual terms – for another pattern of access into the suburb. That raises a difficult question: should that be a totally different one, or should we try to focus both the highway and the rail system on to one central entry point, which would also contain the shops and the community facilities?

The mention of shops raises another hornet's nest. Shaw at Bedford Park could provide one small General Store. Unwin at Brentham was content to provide a modest row of shops. Now the inhabitants will expect a Sainsbury or a Tesco superstore. How do you incorporate that into the urban fabric without destroying all the arcadian quality you are seeking to develop? Would it be possible, as Peter Calthorpe believes, to produce an urban form where large numbers of people would be willing to wheel their shopping home without need for the car? Or will it be necessary to banish the store, just as Unwin did, outside the suburb?

Or would it be possible, perhaps, to square the circle by incorporating a kind of front and back access, one inward-facing and foot-based, the other looking outward and car-based? And if so what does that entail for the pattern of circulation? Can we go back to the principle of segregating foot movement and car movement, first developed at Radburn in New Jersey over 60 years ago, perfected a few years later in the plan for the newtown of Greenbelt in Maryland, outside Washington, DC, and later in the Stockholm suburbs? And can we make that compatible with personal safety – the new planning nightmare of the late twentieth century?

Thames Gateway has curious similarities to a long-lost piece of 1960s planning, from the now-defunct South East Economic Planning Council, which drew on Scandinavian planning practice, described earlier in Chapter 8, to suggest corridor developments along the main transport corridors radiating out from London; it is almost as if one element of that strategy had at last come into being, 30 years later[6] (Figure 46). But it relates also to the new mid-1990s concept from London Transport, described in Chapter 8: the Regional Metro, or Regional TGV.

---

[6] GB South East Economic Planning Council 1967; Hall 1967.

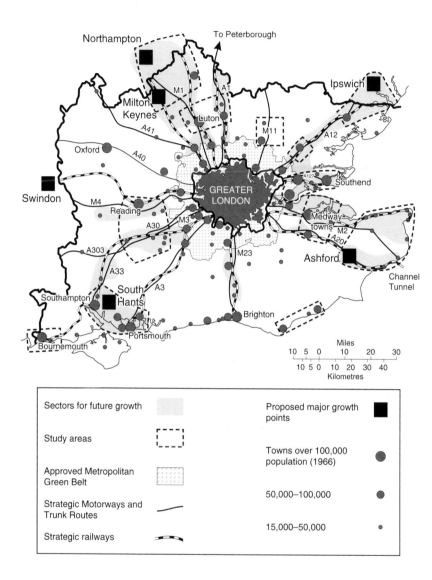

Figure 46 *The South East Planning Council's 1967 Strategy. This proposal, modified in the 1970 Strategic Plan for the South East, proposed discontinuous clustered developments concentrated into the main radial transport corridors. The 1995 Thames Gateway Development Framework finally realises one part of this Strategy.* Source: GB South East Economic Planning Council 1967.

REGIONAL METRO: KEY TO REGIONAL STRATEGY

The planned London Regional Metro is in fact the key to devising a sustainable regional strategy. This system will offer a level of service unknown previously, either in Britain or anywhere else. It will prove another spectacular example of the Colin Clark rule: again, transport will make the city. The impact of the new services will be at least as great as that of the underground extensions of the 1920s and 1930s, on which Lord Ashfield and Frank Pick built modern London. But the spatial impact will be quite different: because of their speed, the new high-speed commuter trains, travelling at 100 m.p.h. (160 km/h) and maybe even faster, will dramatically telescope times to places in the critical range 60–80 miles (100–130 km) from London. For instance, the Channel Tunnel Rail Link will shrink journey times from London to Ashford from 70 minutes to about 40.

It is true that this could encourage long-distance commuting, which would be far from sustainable. But, as suggested above, given that there will be some long-distance commuting, it is preferable that it be on rail rather than on road. And all previous evidence suggests that urban development at distances like these will be relatively self-contained; further, many of the commuters will find local jobs within a few years. So we should base our new settlement strategy on this system.

The key would be to link the regional TGV – for this is what it will be – at key stations to local distributor transit systems, which might be light rail but might equally well be guided busways, such as they have in Adelaide and Essen and now in Leeds, or unguided busways, as in Ottawa. These systems would have a strongly linear form, which might be parallel to the regional TGV or might run at angles away from it; one useful form would connect two TGV stations by an indirect route. Bus transit systems have an advantage over light rail in that they can fan out in dendritic fashion to serve medium-density residential areas more widely spread out from the transit stops, as in Adelaide. In this, however, the important point would be to keep the linear emphasis, which encourages transit use, and at all costs to avoid land uses which encourage cross-trips. Along these transit corridors we would string clusters of mixed-use developments, typically with about 10 000–20 000 residents served by central service concentrations around the transit stations, and further grouped into linear or rectangular units with maximum populations of 200 000–250 000.

SUSTAINABLE DEVELOPMENT CORRIDORS

CITY OF MERCIA

The West Coast Main Line from Euston to Birmingham, Liverpool and Manchester is effectively four-tracked for 82 miles (132 km) from London; four tracks run in parallel from Euston to Roade, 60 miles (97 km) from London, where the Northampton line diverges, but this returns to meet the main line south of Rugby. Shortly to be upgraded at a cost of up to £1 billion by Railtrack in association with

Virgin Trains, to a 140 m.p.h. (225 km/h) standard, this line provides unparalleled potential for creation of sustainable Social City clusters.

Already, it supports one Mark One new town, Hemel Hempstead, 25 miles (40 km) and two of the three Mark Two new towns, Milton Keynes, 50 miles (80 km) and Northampton, 66 miles (106 km) from London. These are separated by open green areas which should be preserved to enhance the identity of the three towns. But between Hemel Hempstead and Milton Keynes, a small new community could be created around the station at Cheddington, 37 miles (59 km) from London, and the small twin towns of Leighton Buzzard and Linslade, 40 miles (64 km) from London, could be expanded.

However, following the basic principles, the major development should occur north of the 50-mile (80-km) point. Milton Keynes should be expanded westwards, to complete the original planned design, thus increasing its population from 143 000 (1991) to its original target of 250 000. Northampton should be expanded northwards in cluster form towards Long Buckby, with reopening of the stations at Church Brampton and Althorp Park, again raising the city's population from 184 000 to about 250 000. However, a substantial open buffer should be left in the fine landscape of the Northampton Uplands between here and Rugby. Rugby itself should be substantially expanded southwards towards Kilsby, with reopening of Kilsby and Crick station, and north-eastwards towards Clifton in Dunsmore so as to round off the town around the station – a development that is fully justifiable by the growth of this area as a key national location for freight logistics, providing a basis for transfer between road and long-distance freight haulage into the European mainland.

The second, and closely associated, rail line is the westernmost branch of Thameslink 2000 between London and Bedford, paralleling the Midland Main Line. This was once the longest stretch of continuous four-track formation in Britain, for 75 miles (121 km) from St Pancras to Glendon South Junction north of Kettering, albeit no longer in operation (and also not electrified) north of Bedford.

The proposal here therefore involves new investment. It involves electrifying, and restoring to four-track operation, the line from Bedford to Glendon, and reopening the passenger service and extending electrification for a further 4 miles (6 km) to Corby. This would support extensive clustered urban development along the line between Wellingborough, 65 miles (105 km) from London, and Corby, 80 miles (130 km) distant. There are also proposals to open a guideway or light rail system connecting Wellingborough and Northampton, and serving the two train stations, a proposal of Wilson and Womersley in their original designation report for Northampton as long ago as 1967, and again under active local examination 30 years later.[7]

Here, again, there are limited possibilities for expansion and new settlements south of the 50-mile (80 km) point. Flitwick and Ampthill, 40 miles (64 km) north,

---

[7] There exists the right-of-way of a railway between the two towns, closed in the 1960s. It runs 2 miles (3 km) south of the proposed guideway, through meadow land that should be preserved as part of a green buffer. It should, however, be reopened to support a cluster of new communities on its south side.

could be expanded to form a clustered development which would involve the reopening of Ampthill station, and further clustered development could occur in the restored brickfield area south of Bedford, necessitating a new station or stations in this area. North of Bedford, the rather fine open countryside of the limestone backslope should be retained as a green buffer, though there could be limited expansions of Oakley–Clapham and of Sharnbrook, both around reopened stations, and a rather larger-scale expansion around a reopened station at Irchester, 63 miles (101 km) from London, which would mark the boundary of more intensive development.

Effectively, the proposal north of this point would be to expand what is already a kind of polycentric conurbation along the Nene Valley. It would form a horseshoe about 40 miles (64 km) in length, running from Rugby at its north-western end, via Northampton, Wellingborough and Kettering, to Corby at the north-eastern end. We propose to call it the *City of Mercia* (Figure 47). Within it, all places would be linked by the two railways and the linking guideway. Along the Thameslink extension, Wellingborough would be further extended to the east, to give a more rounded form around the station (which is presently at a completely eccentric location at the town's eastern edge); there could be limited expansion around a reopened station at Burton Latimer and Finedon, 67–68 miles (108–109 km) from London, and a major expansion of Kettering, 72 miles (116 km) from London. Kettering occupies an important strategic position at the crossing of the new A14 highway from the Midlands to the east coast, and is becoming an extremely attractive location for manufacturing and freight logistics. Finally, the Mark One new town of Corby would be considerably expanded on its south-east side towards Geddington, where the station would be reopened, increasing the town's population from 47 000 (1991) to about 80 000.

CITY OF ANGLIA

North of London, there are the two easterly branches of Thameslink 2000, to Peterborough and to Cambridge and King's Lynn, which really need considering together since they diverge only at Hitchin, 32 miles (51 km) north of King's Cross. This common stem forms the basis of Howard's original Social City, which extends from Hatfield (18 miles, 29 km) via Welwyn Garden City (21 miles, 34 km) and Stevenage (28 miles, 45 km) to Letchworth (35 miles, 56 km) on the Cambridge branch. The East Coast Main Line also has the Mark Three new town of Peterborough, 76 miles (122 km) north of London, currently being massively extended by a new township on old brickfields towards the south.

Here we could start within London, along the Lea Valley, which may have one of the future branches of Thameslink 2000 but is presently served by trains from Liverpool Street via Bishop's Stortford to Cambridge, which runs almost uselessly through a wasteland of industrial sites and sewage farms. Alongside the parkland and lakes of the Lea Valley Regional Park, we could create dramatic walls of high-density housing over the industry: not the monolithic housing that gave high density such a bad name in the 1960s, but that humane version developed by

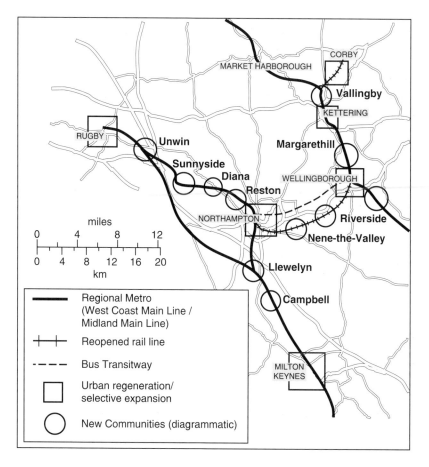

Figure 47 *City of Mercia. Proposal for clustered development along existing and reopened rail lines, plus a Northampton–Wellingborough transit system, incorporating the towns of Rugby, Northampton, Wellingborough, Kettering and Corby.* Source: Peter Hall.

Ralph Erskine in his Scandinavian schemes and in the Byker Wall at Newcastle upon Tyne, which gives every occupant a mini-terrace-garden of his or her own.

We would, of course, preserve the Green Belt, and would not seek more than minor extensions to the four Hertfordshire Garden Cities. But we would develop the corridor between Hitchin and Peterborough, along the main line of Thameslink 2000, by small clusters of new settlements around stations like Sandy, 44 miles (71 km), St Neots (52 miles, 84 km) and Huntingdon (59 miles, 95 km, the end of the four-track formation) and reopened stations at Abbots Ripton (63 miles, 101 km) and Holme (69 miles, 110 km), as well as reopening the station at Yaxley, 73 miles (117 km) to serve the new Peterborough southern townships. Finally Peterborough itself would be expanded westwards towards Castor, thus completing the design of the original 1966 Tom Hancock master plan for the new town, and raising its population from its 1991 total of 138 000 to 200 000.

Along the Cambridge branch, there would be only limited expansion at Royston, 45 miles (72 km) from London, and the general limit on the growth of Cambridge – 58 miles (93 km) via this route or 56 miles (90 km) via the Liverpool Street line – would be maintained. Since the Cambridge area seems certain to expand on the basis of high-technology industry, the general principle – long accepted in local plans, but fiercely contentious in practice – would be to provide for the growth in the form of clustered communities along public transport lines, with a preference for development on the less pressured northern side of the city.

To this end, the key will be to construct a light railway or guided busway on lines long abandoned to passengers, from Sandy to Cambridge and from Cambridge back to Huntingdon, along which could be strung a series of small new settlement units – new villages, in effect. At least one of these would be the Cambourne new village, planned by the South Cambridgeshire district planners near the village of Caxton; others would be in the vicinity of Longstanton and St Ives between Cambridge and Huntingdon. These could be supplemented by similar clusters along the King's Lynn railway north of Cambridge at Waterbeach, Stretham (with a new station) and Ely, setting an effective outer limit on this corridor 70 miles (113 km) from London.

Together, the clusters along the East Coast Main Line, and along the Cambridgeshire transitways, would constitute another polycentric Social City; we propose that it should logically be called the *City of Anglia* (Figure 48).

CITY OF KENT

Finally, we come south and east of London. As already noted, London and Continental Railways' winning bid for the Channel Tunnel Rail Link includes regional TGV services from Rugby and Northampton and Milton Keynes, via St Pancras, to Stratford, Ebbsfleet and Ashford. It is an essential feature of CTRL that it is a dedicated very high-speed link without provision for intermediate stops save for those designated for international service at Stratford, Ebbsfleet and Ashford International. There must therefore be a development gap between Ebbsfleet, 23 miles (37 km) from St Pancras, and Ashford, 56 miles (90 km) distant, and this conveniently corresponds to the very fine landscape of Mid-Kent, the garden of England, which needs preserving against development pressures.

Though there is a domestic branch at Ebbsfleet to the existing North Kent Main Line via Rochester and Sittingbourne to Herne Bay and Margate, this is relatively slow compared with the domestic branches that will exit from the CTRL at Ashford.[8] Even using the CTRL, trains via Rochester will still take 100 minutes to reach Margate, a saving of only about 8 minutes. Ashford can however be reached in 40 minutes, and from there the trains can fan out on the existing electrified rail lines to serve the ring of towns around the East Kent coast: Hythe, Folkestone,

---

[8] A further exit would be possible at the Cheriton marshalling area immediately before the English portal, but is not currently planned.

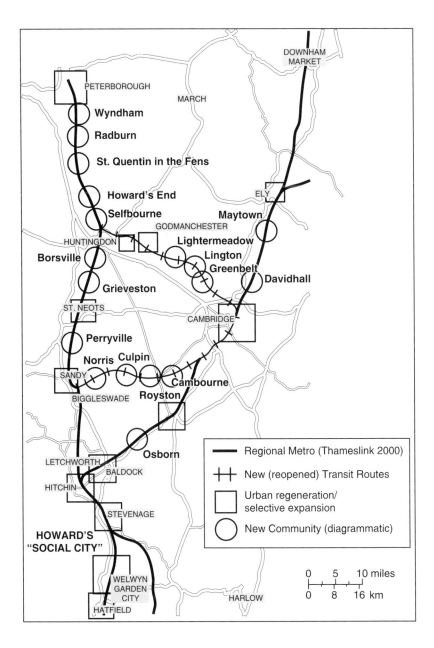

Figure 48 *City of Anglia. Proposal for development clusters along reopened rail rights of way, which could be rail or busway, centred on the cities of Cambridge, Huntingdon and Peterborough and incorporating the new settlements proposed for Cambridgeshire.* Source: Peter Hall.

Dover, Deal, Ramsgate and Margate. Hamstreet south of Ashford could be reached in 49 minutes, Folkestone in 50 minutes, Canterbury in 60 minutes and Ramsgate in 82 minutes. If the existing lines from Ashford could be speeded up – and they are very leisurely compared with the best 125-m.p.h. (200 km/h) on the national network – then these times could be cut by 10 minutes or more. And it might be possible, by reopening a disused line from Canterbury to Whitstable, to provide a dramatically improved service to Whitstable and Herne Bay, with a Parkway station on the new A299 Thanet Way.

The coastal towns which are arrayed almost like a string of beads along this coast – Hythe, Folkestone, Dover, Deal, Sandwich, Ramsgate, Margate, Herne Bay, Whitstable – form in effect another horseshoe-shaped linear conurbation, a little longer than the Nene Valley cluster, being some 50 miles (80 km) long between Hythe and Whitstable. Developed as port or seaside towns, they have suffered serious losses from the gradual erosion of the British holiday industry and finally – in the case of Dover and Folkestone – from the Channel Tunnel; they need a new role. By exploiting their new accessibility to London – one of the most dramatic, in time terms, ever achieved in one step – they could begin to build an economic base of their own, as has occurred progressively in that more favoured sector of South East England, north and west of London. The branch of this system which loops from Dover via Deal and Sandwich to Ramsgate connects also with inland settlements such as Patrick Abercrombie's eight planned 1920s new towns, one of which – at Aylesham – actually got started, and most of which are next to the railway lines.[9]

There is one further element: south of Ashford is the highly anomalous branch line via Rye to Hastings, one of the very few lines in the entire system south of London that is still not electrified. For 8 miles (13 km) south of Ashford, Kent County Council have established what is in effect a growth corridor via Hamstreet to Appledore. Electrified and linked directly to the CTRL, the Hastings line could convey extraordinary accessibility to this corridor. At Rye the line enters the South Downs Area of Outstanding Natural Beauty, where the development potential is minimal. But a freight branch, which runs out from Appledore to Dungeness power station, could provide an extension of electric traction, permitting clustered development at Brookland, Lydd, Lydd-on-Sea and Romney.

Together, the development and regeneration of the East Kent coastal towns, together with development south of Ashford into Dungeness, would constitute a third great polycentric city region based on high-speed rail linkages; we propose that it be called the *City of Kent* (Figure 49).

IMPLEMENTING THE STRATEGIC DESIGN

The West Coast Main Line, the three branches of Thameslink 2000 and the Channel Tunnel Rail Link should without doubt form the first three development

[9]   Dix 1978, 337–339. An old coalfield line, reopened, could provide an additional link.

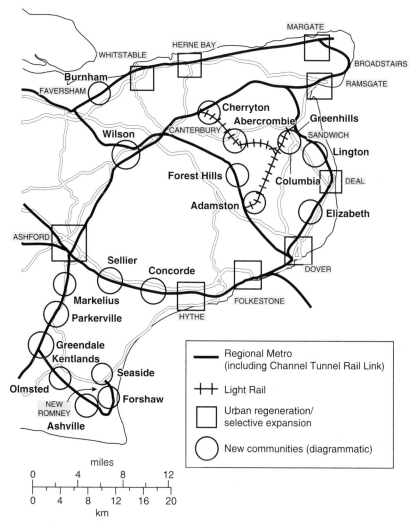

Figure 49 *City of Kent. Proposal for major development clusters served by domestic services on the Channel Tunnel Rail Link, centred on Ashford but incorporating the regeneration and extension of the depressed port and seaside towns of East Kent, and the revival of the Kent coalfield.* Source: Peter Hall.

corridors, for the simple reason that they constitute the initial lines of the new regional TGV system that are certain to come into existence before 2005. Further, the three Social Cities that we propose on these corridors are precisely in the right places from a regional planning point of view. The South Midlands are a logical place for major development just beyond the zone of maximum pressure at the edge of the South East, while East Kent is in urgent need of development. When CrossRail is built, probably about 2010, these three corridors could be supplemented by another corridor to Colchester, Ipswich and the Haven Ports of Harwich and Felixstowe.

The precise timing and balance will need to be refined. But one point is crucial to the political achievement of the project, and that is to balance the development emphasis along the corridors, with a much greater degree of restraint on development outside them. The development corridors should be the way of guaranteeing the tranquillity of the remaining areas. Logically, in the wide rural wedges which lie between and away from the main road and rail routes, the countryside and the traditional patterns of village settlement and village life could be preserved and cherished. Similarly, even within the corridors, it is vital to stress that all the land would not by any means be continuously urbanised; that is quite contrary to all the principles on which the development is postulated. Substantial areas of exceptionally high agricultural value or landscape value would continue to be protected even here: the Mid-Kent belt between the Medway towns and Ashford provides a prime example.

We are going to be forced to revisit the housing question, not because of the existence of the Victorian slums which so shocked Howard and his more thoughtful contemporaries, but because we face a huge and yawning gap between demand and provision: a gap that is soon going to bring housing back into centre stage of the political theatre, from which politicians of all complexions thought it had been banished. And, when that happens, planning will come back with it, hand in hand. We face a very interesting few years, in which a new government has to be persuaded of the wisdom of this argument. It is oddly like the situation in which F.J. Osborn found himself in 1946, when he had to persuade Lewis Silkin, planning minister in Attlee's government, of the virtues of new towns; or again in 1964, when he similarly had to educate Richard Crossman. History does repeat itself, though not always in identical clothing.

CHAPTER 10

MAKING IT HAPPEN[1]

The kind of programme outlined in Chapter 9 will not be achieved easily or automatically; it will require deliberate action at different levels: national, regional and local. This chapter looks at the agencies and mechanisms that will be needed.

There are five major challenges:

1. To achieve large-scale sustainable, balanced packages of regeneration and redevelopment in each region, and in particular in the pressured regions of southern and midland England.
2. To develop appropriate agencies and mechanisms to bring the land forward in the right amounts and at the right times for this purpose.
3. To do so without putting excessive claims on the public purse. Traditional agencies that proved effective in past decades, like the New Town Development Corporations and their successors the Urban Development Corporations, have become suspect because of the claims they appear to make on public expenditure in an era of fiscal stringency. Private enterprise is ready and waiting to build the new homes, so long as it can find buyers; private finance is available to provide the necessary funds for owner-occupied housing, but it will not itself be able to find the necessary land for the job. The problem is how to marry private finance and enterprise with strategic planning and with development and funding procedures, so as to put the new homes in the right places, thus to produce a pattern of development that is convenient, efficient, equitable and above all sustainable.
4. Specifically, to find funds to pay for the necessary infrastructure. For over half a century, debate has raged in Britain over the right way to recoup the share of the profits from land development that rightly belongs to the community, since public agencies have had to provide much of the physical and social infrastructure, and since the land value arises in large measure through the grant of planning permission. What has eluded us, all this time, is to find a way of capturing this added value that is effective, efficient in operation, and politically acceptable enough to be stable over time.

---

[1]   This chapter is substantially based on the TCPA report *Building the New Britain* (Town and County Planning Association 1997), produced by a working party under the chairmanship of Peter Hall, and on the discussion of that report at a TCPA conference in October 1997.

5. To provide an adequate mixture of social housing in each area and in each major development. For the foreseeable future, a supply of affordable housing will be needed for a considerable remaining minority of the people.

These problems have plagued us ever since the historic 1947 Town and Country Planning Act, the origin of the planning system that has successfully guided the development of British towns and countryside for over half a century. But the scale of the current problem gives it a new urgency.

BUILDING ON PAST LESSONS: THREE ATTEMPTS – THREE FAILURES?

Politicians, especially at major junctures like this one, do not normally pay much attention to history. But in this matter, they should. Three times in the last 50 years, Labour governments have legislated on land. Each time, they had the same twin objectives, well set out in the 1965 White Paper that preceded the second of these attempts: "to secure that the right land is available at the right time for the implementation of national, regional and local plans", and "to secure that a substantial part of the development value created by the community returns to the community and that the burden of the cost of land for essential purposes is reduced".[2] Each time, the legislation they drafted ran into well-nigh insuperable difficulties; each time, a succeeding Conservative government more or less promptly repealed all or part of the legislation, bringing us back – but not always full circle – to the beginning. We need to start by asking why. And to start answering this, we need to revisit each of these three crucial periods of recent history.

First, *1945–51*: the 1947 Town and Country Planning Act effectively national-ised development rights and their associated land values, on the basis that all betterment was created by the community and therefore belonged to it; payments for lost development value would be compensated from a £350 million fund, not as of right but to alleviate hardship. Accordingly, under powers given by the Act a Central Land Board levied a 100 per cent betterment charge on all gains in value arising from development. The scheme effectively began operation in 1949. However, only two years later, in 1951, the new Conservative government argued that this would act as a brake on private development, and put it on hold; in 1954 they terminated Development Charge.

But, very importantly, they did not try to repeal the 1947 Act itself. Land development rights remained nationalised, and have remained so to this day. The anomaly was that there was now no logic in the system: landowners could again make huge profits from their land, so long as they could get planning permission to develop it. And in 1959, the same principle was extended to compulsory

---

[2] Quoted in Cullingworth and Nadin 1997, 140.

purchase by public authorities. However, there was an important exception: in Section 6 of the 1961 Land Compensation Act, under what came to be known as the Pointe Gourde system, "the effects of the scheme on values are to be disregarded". In other words, compensation for land acquisition should not include any value that resulted from the very scheme that gives rise to the acquisition. In the first column of the schedule to the Act, it was specifically provided that a new town development corporation would not have to pay for value created by the designation of extension of a new town. However, subsequent case law has established that if hypothetical planning permission would have been granted for something akin to the scheme, then the value of that would have to be taken into account.

The 1947 Act has been the foundation of our planning system for almost exactly half a century. It has allowed local planning authorities to plan for the best use of land, without a concern that this would entail huge compensation claims for lost development rights. They have used these powers to plan for the supply of housing land, generally keeping a five-year supply ready for development, and revising their plans regularly. Though there have been major debates about whether they were providing enough land, and about the possible effects of restricted supply on land and housing prices, the system has worked relatively well. But with one exception: the system could not guarantee that sufficient land was made available for affordable housing. Only recently has the government encouraged local authorities to make specific allocations for this purpose.

Nor did the Conservative governments of 1951–64 repeal the Labour government's 1946 New Towns Act, under which a start had already been made on eight new towns around London and others in Wales, the North East and Scotland. True, they were reluctant to start further new towns (except for Cumbernauld in Scotland, in 1955); they preferred to work through the Town Development Act of 1952 (actually prepared under the outgoing Labour government), under which city authorities could strike deals with rural towns to accommodate overspill, with the government contributing money for essential infrastructure. But even they were compelled by rising demographic pressures to revive the new towns programme from 1961 onwards.

Second, *1964–70*: the 1967 Land Commission Act established a Land Commission, with wide powers to buy land voluntarily or compulsorily. It also initiated a betterment levy, to be deducted by the Commission when it bought land, but also payable on realisation of the value by private owners, so ensuring equal treatment for both private and public purchases; unfortunately, the fact that the levy was payable on realisation (not on disposal) meant that individuals were often faced with bills that they could not pay. This was initially set at a reasonable 40 per cent, to give owners an incentive to sell, but it was planned to rise to 45 and then 50 per cent (which never happened). The Land Commission was abolished by the Conservatives in 1971, because they said that it had failed to reduce prices, ease land supply problems or assist the efficiency of the market. In fact global forces, in the form of a major fiscal crisis, had intervened. Just as on the previous occasion, there had simply been insufficient time to test whether it could have worked.

The Labour government of 1964–70, however, did something else. Following a lead set by the outgoing Conservative government, which had published *The South East Study* just before the 1964 election, it commissioned work on a major strategic regional plan for South East England and parallel studies in other regions.[3] These established long-term strategies for the development of major growth zones in some places and widespread protection of rural land in other areas. Though these plans were not always rigorously followed, there is little doubt that, by providing strong guidelines to local authorities, they helped avert anarchy at a time of big demographic pressures. Later on, in 1983, the first Thatcher government was to deride strategic regional planning as a fashion from the 1960s, the time for which had passed; but before long, the third Thatcher government found itself establishing the system of regional planning advice and guidance that persists to the present day.

Third, *1974–79*: in 1975, the Community Land Act proposed progressive extension of public ownership of development land – in England and Scotland, through the agency of the local authorities, but in Wales (with its smaller local authorities) an *ad hoc* agency, the Land Authority for Wales, was created. The ultimate basis for purchase was to be current use value, with sales at market value; provisionally, there would be a development land tax.[4] The Act was effectively killed by the deepening fiscal crisis of 1978; by April 1979, just before it was wound up by the Conservative government, the Community Land Account was £33 million in deficit.[5] But, in terminating this third attempt, there was one remarkable detail: the first Thatcher government retained the Development Land Tax, effectively absorbing it in Capital Gains Tax, because there was widespread consent – among landowners and developers – that it was reasonable. It was retained, at 40 per cent, until in a remarkable act of political subterfuge it was quietly rescinded by Nigel Lawson in his 1985 budget.

LEARNING THE LESSONS

Why did each of these three attempts fail? One good reason was timing. Each one needed legislation, invariably two years into the term of a government, and another one or two years to bring it into operation; by that time, another general election was looming, followed by a change of government or (in 1950) by a hung parliament. The lesson is clear: any such procedure is almost doomed to destruction, because it is politically unstable. An effective solution is an early solution, brought into operation ideally within the first year, at very least in the second, of a reforming administration. That is the approach we follow in the package we now set out.

---

[3] GB Ministry of Housing 1964; GB South East Joint Planning Team 1970.
[4] Cullingworth and Nadin 1997, 136–143.
[5] Cox 1984, 187.

SOLUTIONS: (1) PLANNING FOR REGIONAL HOUSING NEEDS

Since 1979, successive governments have presumed that the statutory planning system would allocate enough land to meet the nation's housing needs. The "market" would then develop a sufficient supply of housing for sale. Generally, the market has brought sufficient land forward for private housing – though with great stresses and strains in the more pressured parts of the country. We know that planning restraints, both in these areas and in other parts of the country, have raised the cost of development land and hence the cost of housing. But for social housing there is an additional and real problem. Most of the land owned by local authorities, and development bodies, was systematically sold for private and social housing; these public land banks were not replenished. Housing associations were expected to buy land at open market value for social housing.

As a result, most local authorities do not own sufficient land suitable for social housing. The only significant housing land holdings in the public sector are held by the Commission for the New Towns. These are the undeveloped parts of the Mark Three new towns. Moreover, every effort has been made to sell the sites which were assembled by Urban Development Corporations for private housing prior to the winding-up of these bodies. Those sites which the UDCs cannot sell will be transferred to English Partnerships or the Commission for the New Towns. But this will be insufficient to remedy the real shortage of land to meet the backlog of social housing.

As a result, the current Labour government faces a choice: it must decide whether to rely completely on the planning system, and the private market, to bring housing land forward for development, or whether to adopt a more proactive approach. The New Town Development Corporations, the Land Authority for Wales, the Urban Development Corporations, and schemes promoted under the Town Development Act are prime examples of such a proactive approach. Notably, they have been successful in producing communities with a balanced mix of social and private housing. We believe that the government must find a proactive approach to bringing land forward for housing which supplements the private land market.

There are, we believe, three parts to the solution, and they are necessarily linked.

The first is to provide a more effective planning mechanism, at national and regional levels, to bring land on stream in the right places and at the right times to achieve balanced regional packages of sustainable urban brownfield regeneration and sustainable first-time development.

The second, following this, is to create effective agencies to provide the mechanisms for making the land available and ready for development in these places. This assumes that the great bulk of development, including regeneration and new build, will be carried out by the private sector; the key is to provide the land, fully serviced and ready for development.

The third is to find a way of paying for all this without making an undue claim on the public purse; indeed, to do so while providing a steadily accelerating income stream both to the Treasury and to local authorities.

The first essential is a strong framework of national and regional guidance. The government must first produce a definitive national Planning Policy Guidance note on the ways in which it proposes to deal with the challenge of the 4.4 million additional households. It must, in particular, set out the principles that should guide regional and local bodies in developing appropriate bundles of urban regeneration and new development. But necessarily, much of the work will be done at regional level; for the precise balance to be struck will depend greatly on the precise geography of each region. London and the South East report severe constraints on the availability of brownfield land, as do the South West and Northern regions, while in the North West much of the need could be met within the conurbations.[6]

Therefore, the first key step is to produce stronger, more detailed, regional guidance. It will need not only to indicate broad county housing targets, as now, but to develop broad regional planning strategies that will go across county lines (and even, in the case of the South East–East Midlands, across regional lines). It would do this in a positive and proactive way (not by pro-rata allocations as has been all too common in recent guidance) by defining Priority Areas for Regeneration and Development (PARDs): as a first step, Government Regional Offices would take the broad regional requirement, set by the Department of the Environment, Transport and the Regions headquarters, would compare this with provisions in structure and local plans, and would then, in a regional conference convened by the Secretary of State, establish the boundaries of PARDs with an indication of the broad location of new settlements and town expansion schemes.

In the longer run, these plans should and will come from elected regional authorities. But it will take years, almost certainly the lifetime of more than one parliament, to extend this principle to the entire country. So the existing mechanism, of Regional Standing Conferences to proffer advice and Government Regional Offices (or Regional Development Agencies) to produce regional guidance, will continue, and the latter should be made much more prescriptive.

As the government statement recognises, the counties and unitary authorities need to be more actively involved in this process, for their officers have the detailed expertise. But that expertise has been very seriously eroded by expenditure cuts in recent years. It will also be very important that the entire process is not compromised by local NIMBY pressures, which are all too likely to be present. We therefore think that it would be useful to second county and unitary planning officials to the Regional Offices to participate in production of Regional Guidance, assisted where necessary by independent consultants, but that the established distinction between Advice and Guidance should be maintained. (The government, in the event, has gone further than this, suggesting that they be collapsed into one process.) We also think it essential that, within the new integrated structure of Environment and Transport, Regional Ministers be established with responsibility for each of the Regional Offices, directly charged as a principal responsibility with the organisation of the proposed regional conferences and the

---

[6] Breheny and Hall 1996a.

subsequent delivery of agreed Regional Guidance. This is the short-term prescription; eventually, Regional Guidance should (as part of the legislation creating Regional Authorities) be given statutory force as a Regional Plan.

The resultant Regional Guidance would be more detailed than at present, having the character of a Regional Structural Plan. This will then need to be turned into more detailed statutory local plans. In many cases these will need to run across district – and even, on occasions, county – boundaries, and will in many cases cover both areas now subject to Unitary Development Plans and areas subject to the two-tier Structure/District Plan system. Given the fact that structural guidance will already have been issued, these would be local plans (or Unitary Development Plans Part Two). The Secretary of State already has powers to request local planning authorities to produce joint structure plans where appropriate, if necessary calling upon Government Regional Offices for this purpose; we think that the same principle should operate here. However, this very much depends on the precise agency chosen to carry out the development – the point to which we turn in the next section of the chapter.

The government's flurry of statements early in 1998,[7] *Modernising Planning* and *The Future of Regional Planning Guidance*, followed by *Planning for the Communities of the Future*, go only some way to resolving these issues: the big decisions, that would fundamentally reshape the 1947 planning system, are still several years away.

Modernising Planning says that there must be more specific national planning guidance – a real innovation, if it gets followed through. It also demands that local authorities produce new streamlined plans, shorter and clearer, and with targeted consultation. And it says that the system of planning obligations should be made more predictable and transparent. We could find little to disagree with here. Indeed, the statement seems to echo some of the ideas the TCPA put to Nick Raynsford immediately after the 1997 election. With a plan-led system in a highly dynamic world, it must be right to speed up the plan-making process. With public expenditure squeezed, it must similarly be clever to extract the maximum possible planning gain on the maximum number of occasions.

Modernising Planning also trails *The Future of Regional Planning Guidance*, which sets out the central proposals in more detail. And here, the government reveals itself to be in something of a quandary. It needs a stronger but also a more regionally based system than the highly centralised one developed over the past decade by Conservative governments. But it has delayed the arrival of elected regional government for at least five years. It still proposes Regional Development Agencies, but they are non-elected. So it proposes what it calls a hybrid solution: instead of the double two-step of recent years, consultative advice/final advice/ consultative guidance/final guidance, it suggests that regional conferences of local authorities go through two stages in association with government offices, with "consultation" and "modification", and with a kind of public inquiry.

---

[7] GB Department of the Environment, Transport and the Regions 1998a, 1998b; GB Deputy Prime Minister 1998.

It says more even than that: it wants the new strategies to be "spatial", not merely dealing with land use, but embracing other aspects like transport and the environment, and hence providing a strong steer to other agencies including private ones. It proposes the development of sub-regional strategies for awkward areas like cross-border zones which are under pressure – as for instance where SERPLAN abuts on the South West, East Midlands and Eastern regions. And it wants to see an environmental appraisal as part of each structure, evolving rapidly into a "sustainability appraisal" covering different aspects of sustainability – economic and social as well as environmental.

Again, we can only applaud – for the TCPA has been saying the same things. Producing regional guidance is a devilishly difficult business, because it means trading off so many different legitimate interests: urban NIMBYs as well as rural ones, future residents as well as present ones, and even households as yet unformed as well as well-organised interest groups. And, because the government have backed away from elected regional authorities for the time being, it leaves the regional conferences – federations of local interests – to propose the critical strategies: not exactly a recipe for strong or imaginative thinking.

The February 1998 statement, *Planning for the Communities of the Future*, came as the response to the immediate expression of this conflict of interest: the dilemma of providing for the 4.4 million households predicted by 2016, while grappling with an extraordinarily well-organised and well-financed campaign to save the countryside. Basically, as noticed in Chapter 7, it shifts back to the previous target set by John Gummer: 60 per cent of all new homes, nationally, to go on to brownfield land – but only over a 10-year period, and with regional variations. And, following the January statement, it says that regional strategies should be set by the conferences of local authorities – albeit with a certain amount of arm-twisting by the regional authorities.

In other words, it could be argued, the government has dumped the problem in the local authorities' lap: it seems to be saying that they should fight it out among themselves and somehow reach agreement, But it does not leave everything to chance. The new White Paper falls back on the formula of the earlier statement: yes, the regional conferences will be given every encouragement to reach agreement on the right strategy, but if they mess around too long, the Government Regional Office will always be there to encourage them.

Here, the White Paper's detailed language is more interesting than the lofty aspirations: "The relevant GOR would play an active role in drawing up draft advice;" and, "In exceptional circumstances if he [the Secretary of State] were minded not to accept the Chair's recommendations on the housing figures, or any other aspects of the draft RPG, he would publish draft modifications with a statement of reasons;" and, finally, "Publication of the RPG would remain the responsibility of the Secretary of State." Moreover, once published, "the presumption would be that the housing figures would be reflected in structure plans and UDPs". Housing figures could be discussed in Examinations in Public (EIPs) or inquiries, but focusing on locations.

Clearly, this is a long way from regional autonomy Barcelona or Bremen style; it is lightly-modified old-style centralism. And the reason must be blindingly obvious:

in this transitional period, the government does not feel safe in handing over responsibility, lock, stock and barrel, to the regional conferences. That way lies NIMBY pulling-up of drawbridges, and that is something no government could contemplate with equanimity.

Centralism, logically, appears in the guidance on tackling the housing challenge: the national target is 60 percent brownfield, but this will be allowed to vary from region to region. In 1991–93 the actual percentage, in terms of dwelling units built on brownfield, ranged from 87 per cent in London and 69 per cent on Merseyside, to 37 per cent in the South West and 35 per cent in the East Midlands. The implication is pretty obvious: the conurbations, above all London, will be encouraged to go above 60 per cent (and they could hardly do other) but the shire county regions may well go lower without any government reprimand.

The shires will not of course be minded to do so. The question is first whether this will defuse the immediate crisis presented by the countryside revolt, and second whether there is a long-term way of putting regional strategic planning in the hands of regional authorities. It is not going to be easy, because they may prove only as resolute as their most conservative-minded members, and because some of the strongest pressures are now being felt on the regional boundaries – as between Newbury and Swindon, Milton Keynes and Northampton, Bedford and Wellingborough, or Stansted and Cambridge. Sub-regional strategies must be the answer in such places, but they would not easily conform to a system of regional government.

Given a modicum of luck, the government has given itself a four-year breathing space, but some hard thinking will be needed. We now return to our own suggestions.

SOLUTIONS: (2) FROM PLANNING TO IMPLEMENTATION

Despite the historical record of legislation and counter-legislation, it is remarkable that there still exist wide powers to create land development agencies, while in some cases the agencies themselves continue in business. We need to review these and then to ask which one of them, or which combination of them, could best undertake the massive job of assisting the private housebuilder to do the job of providing the 4.4 million new homes required.

THE NEW TOWN DEVELOPMENT CORPORATIONS

The New Towns Act of 1946, still on the statute book (Section 1 of the 1981 New Towns Act), provided that the new towns were to be planned and built through the agency of development corporations, generally one for each town, to be set up by the Secretary of State wherever he or she was satisfied "that it is expedient in the national interest" to do so. They were given wide powers: "to acquire, hold, manage and dispose of land and other property, to carry out building and other operations, to provide water, electricity, gas, sewerage and other services, to carry on any business or undertaking in or for the purpose of the new town, and generally to do anything necessary or expedient for the purposes of the new town or for the purposes incidental to do so". This meant a delicate relationship with the local authority, which retained planning powers. The "partnership" new towns of the 1960s, which were created around already established towns like Northampton, Peterborough and Warrington, and where local councillors – four in each case – were given a major role on the development corporations, were only partial exceptions.[8] In Northampton the partnership is said to have worked well, though the town insisted on retaining the planning powers within the old built-up perimeter, so that there is a lack of relationship between the old and the new areas.[9] In Peterborough, former General Manager Wyndham Thomas recalls, it took almost daily work to keep the relationships smooth and co-operative.

The development corporations were funded by Treasury loan through the Consolidated Fund. Down to the mid-1980s, £3625 million had been supplied in the form of loans repayable over 60 years, with interest payable at the date on which the advance was made; there was no subsidy through a concessionary rate, except for the fact that the Treasury's position allowed it to borrow at slightly less than market rates. The Treasury insisted that every proposal must be subject to the Secretary of State's approval, which meant a great deal of detailed vetting by the Department of the Environment.

The procedure for setting up a new town development corporation is necessarily long and complex. The Secretary of State for the Environment, Transport and the Regions may designate an area of land for development as a new town by a development corporation. He or she must be satisfied that this is expedient in the national interest. He or she can make a Statutory Instrument for this purpose, but this can be annulled by a resolution of either House of Parliament if a county planning authority objects. A detailed procedure provides for a public inquiry into the order. Only after designation may the Secretary of State establish a development corporation. The whole process can take several years.

The Commission for the New Towns

In the New Towns Act 1959, the government of the day provided that on winding up the Development Corporations, ownership and management should pass to an independent public body: the Commission for the New Towns. CNT is a permanent body, charged with the duty to "maintain and enhance the value of the land held by them and the return obtained by them from it" while having "regard for the purpose for which the town was developed and to the convenience and welfare of the persons, residing, working or carrying on business there".

Then, in 1976, the New Towns (Amendment) Act provided for the transfer of housing and all related assets (i.e. land and buildings) to local authorities; this

---

[8]   Cullingworth 1985, 248.
[9]   Anon. 1996, 304.

happened in April 1978, though much of this housing has now passed into private ownership.[10] Nonetheless, in 1998 the Commission still had substantial commercial assets in some 19 English new towns, which it advertises and promotes vigorously. Its powers – spelled out in Section 36 of the 1981 Act – are, however, limited. Unlike the development corporations, it is primarily concerned with land disposal rather than land assembly and development. Its powers in these areas are essentially to achieve its main purposes. Legal advice is that any extension of the Commission's powers to allow it to become active in direct development would need additional primary legislation.

THE URBAN DEVELOPMENT CORPORATIONS

The Urban Development Corporations were created by the Local Government, Planning and Land Act of 1980. Michael Heseltine himself was on record as saying that the UDCs were consciously modelled on the new town development corporations; the intention, in his own words, was to "create new towns in old cities". This might be thought ironic, given that the NTDCs were the creation of a radical left-wing government and the UDCs the creation of an equally radical right-wing government, but in essence the logic was the same, to create public corporations independent of the local authorities in whose areas they were located, and with freedom to acquire and develop land. Exactly like the NTDCs, UDCs were accountable only to Parliament; they were boards appointed by the Secretary of State, not by elected representatives.

The UDCs were, however, given much wider powers than the New Town Development Corporations had ever had: the Secretary of State for the Environment could simply make an order authorising that land held by local authorities, statutory undertakers, or other public bodies be vested in the UDC. This particularly applied to the large areas of land held by major undertakers such as the Port of London Authority, the Mersey Docks and Harbour Board or the former Gas Board, all of which was transferred automatically without right of appeal. In fact, the great bulk of land passing to the London Docklands Development Corporation was vested in this way.

The most important extension of powers, however, was that, for so long as they remained in existence, the UDCs would supersede the local authorities in their areas as development control authorities: they were to decide on planning applications and determine what should and should not receive planning consent. They must "have regard to the provisions of the development plan", whatever that meant, but if there was a difference with the local planning authority, the matter would be referred to the Secretary of State, who could grant planning permission through a Special Development Order (which would require Parliamentary approval); the general effect was that the UDCs subsumed the planning powers. By this means they could grant development permissions without any consultation

---

[10] Cullingworth 1985, 255.

or inquiries. Interestingly, in Wales the Cardiff Bay Development Corporation – responsible for a regeneration scheme second only in scale to London Docklands – has these powers but chooses not to invoke them; it voluntarily works in co-operation with the two local authorities in its area, which retain all development control powers.

The essential concept underlying the UDCs was the American one of leverage. The sweeping powers granted to the UDCs to own and acquire land, build factories, and invest in infrastructure and environmental improvements – many of which they did not use in practice – were in order to attract private funding in industrial, commercial and residential development. The central idea was that a certain amount of public investment could effectively kick-start the process of urban regeneration, attracting a much bigger volume of private funds.

This is not the place to review the achievements or the limitations of the UDCs; that is a matter for history, for they are being wound up and their assets transferred to the CNT. The main question now is the disposal of the remaining, still under-developed, land in the UDC areas. As with the new town land, CNT's remit is limited. It is not a development agency, and its powers to provide finance for the development of difficult sites are limited. It will therefore inevitably find itself working in a close relationship with English Partnerships.

ENGLISH PARTNERSHIPS

English Partnerships was created in 1994, the result of Michael Heseltine's arguments for an English Development Corporation, a body which would parallel the Scottish Development Agency (now Scottish Enterprise) and the parallel Welsh Development Agency, bodies with long track records of economic regeneration in their own parts of Britain. The central idea was that of a kind of roving development corporation which would work nationally but might focus somewhat on major schemes at certain times (as with the SDA's promotion of GEAR, the Glasgow Eastern Area Renewal scheme, in the late 1970s). It was significant that for the very large Thames Gateway regeneration scheme, representing an extension of London Docklands on a much larger scale and extending for some 40 miles eastwards down the Thames, no UDC was proposed; rather, English Partnerships were expected to play a major role through the provision of funds for land preparation.

EP's own description of itself is "principally as an enabler, working in partner-ship with the public, private and voluntary sectors, supporting land reclamation, property development and the creation of strategic development packages for employment, housing, recreation and green space". Since formation in April 1994, it has engaged in a £460 million development programme that has attracted £880 million of private finance and reclaimed 3220 hectares of vacant and derelict land as well as assisting the creation of over 1.5 million square metres of industrial and commercial floorspace, and facilitating the development of 7850 housing units.

There seems to be no doubt that English Partnerships is the logical agency to perform the job of assisting local authorities with funding and expertise for the redevelopment of difficult brownfield sites. The problem is that its limited funds are nowhere near enough to meet the scale of the problem; that is already clear, for instance, at Barking Reach, the first major development site in Thames Gateway, where development is stalled pending provision of funds for waste clearance and undergrounding of power lines. If any government is serious about the commitment to 50 per cent brownfield development, let alone 60 per cent, the key can only lie in providing much more specific financing for restoration of such difficult sites, thus providing a level playing field *vis-à-vis* easier first-time development sites.

THE LAND AUTHORITY FOR WALES

There is a rather remarkable survivor of the 1975 Act: the Land Authority for Wales. It was established with the aid of a £20 million start-up Treasury guarantee (which would translate into about £40 million at 1998 prices), sufficient to allow it to work toward a balanced budget within a few years. Until 1985, when Development Land Tax was abolished, it bought land on the open market net of DLT; it still buys it at prices which discount the effect of any planning scheme (the Pointe Gourde principle). It then holds, services and develops the land. It only brings land on to the development market when it believes it is socially and economically necessary for private and public sector house building and industrial and commercial development. Sales normally take place at full market prices to the private sector, accumulating a surplus for future activities. It has been argued that the LAW is successful because it is divorced from local government; it has its own staff of planners, valuers and surveyors, who pursue their activities single-mindedly. Also, while it consults with local authorities, as an autonomous agency it has escaped the recalcitrance associated with political prejudice that colours the work of many local authorities, both Conservative and Labour.[11] It is interesting that Mrs Thatcher allowed it to continue, but apparently without recognising its potential usefulness elsewhere.

WHICH AGENCY?

There seem to be four realistic options. All would require a strong framework of Regional Guidance and the production of Joint Structure or Unitary Development Plans, if necessary implemented by the Secretary of State for the Environment and Transport through appointment of consultants by the Government Regional Offices.

1. *CNT as Land Authority:* The Commission for the New Towns would be renamed English Land, and would be given – possibly through an Order by the

---

[11]   Cox 1984, 200.

Secretary of State for the Environment, Transport and the Regions, but more probably through legislation – additional powers to acquire and dispose of land, and to provide infrastructure, within the areas specified by the Joint Development Plans. English Land would not develop this land itself, but would pass it over for development under covenant to private developers/builders, on the model of Peterborough Southern Townships.

2. *Merger of CNT and EP*: The same as (1) above, except that the Commission for the New Towns and English Partnerships would merge under the name English Land. The new body would have the extended CNT powers described above, plus all the present powers of English Partnerships to assist regeneration. This would have the merit that the operations of English Land could be managed to provide further aid to regeneration of difficult brownfield sites.

3. *Regional Land Authorities*: As (2) above, but the merged agencies would be regionalised (as indeed is the case for the two bodies now), and would act under the aegis of the government's proposed regional development agencies, as their land development arms.

4. *Sub-Regional New Town Development Corporations*: Within the PARDs (see earlier in this chapter), New Town Development Corporations could be designated to operate over discontinuous development zones or corridors; the parallel would be with the original plans in the 1960s for the Mid-Wales linear new town, or the Central Lancashire new town embracing Preston, Leyland and Chorley. Within these areas, the corporations would exercise traditional new town powers. They would work mainly in association with private developers, but would build some assisted housing themselves as well as making a proportion of their land available to housing associations.

It seems clear that (1) is the easiest, simplest and fastest, and (4) the slowest and most complex. The choice must depend on a political judgement as to the urgency of the problem, and also as to which agency would be most effective. We have no wish to multiply bureaucracy; the aim must be to create an agency only where it is absolutely necessary to help private enterprise do the job it knows best.

On balance, the TCPA view is that the easiest and most effective solution would be to extend the relevant functions of the Commission for the New Towns and of English Partnerships to create English regional equivalents of the Land Authority for Wales,[12] which would work as independent arms of the new Regional Development Agencies; it is a moot point whether they should be linked nationally, under the title English Land. They would work with the new Regional Agencies to support the efforts of local authorities to secure urban regeneration, major town expansion schemes and new settlements. Their activities would enable the increased net proceeds from the disposal of assets to be re-invested – as CNT assets are now invested – in urban regeneration, town expansion schemes and new settlements.

---

[12] The government's latest proposals for regional devolution in Wales appear likely to merge the Land Authority for Wales into a new Welsh Development Agency.

The new agencies would buy development land on the open market net of impact fees, and net also of any special Capital Gains Tax (see below), representing one half of the additional value attributable to a planning scheme on the land. In some cases, when invited to do so by a local authority, they would have to use compulsory purchase powers. They should be empowered to transfer some profits from new development to the preparation of difficult brownfield sites.

When land is being used for social housing, and other community benefits, the right way will be to apply the same principle as with the New Towns Act: the landowner should receive no increase in value that is due to the existence of the scheme itself. This principle is easier to apply in major developments of the new town type; logically, much of the necessary social housing should be in mixed developments within such major units.

The creation of these regional land authorities might not require primary legislation, but possibly could be established through an Order under existing legislation; if not, they could easily be established as part of the proposed legislation establishing the Regional Development Agencies. The agencies would then release land for private developers to carry out the actual development. But, like CNT in the new towns today, they would do so through covenants containing design briefs. Experience in Milton Keynes and elsewhere is that these covenants have produced extraordinarily high levels of urban design and ambience, which are reflected in higher-than-average sales values; but that they can readily be combined with integrated provision of social housing, without any of the resistance often found in residential neighbourhoods. This we believe would be a central advantage of the plan: it could guarantee development, over the next 20 years, of an exceptionally high-quality urban environment both on new-development and on brownfield sites.

But, as already explained, we believe that in designated PARDs the new towns development mechanism should apply. The simplest way to achieve this would be for the Regional Land Agencies to operate as Development Corporations under powers that stem from the 1946 Act. They would operate in the same way as elsewhere, through design briefs, but here they would have greater overall responsibility, since all the land within the designated PARD would be vested in them, for early release to private developers. They would work in close partnership with the private sector to this end, and would draw on private as well as a modicum of public capital to finance land assembly and infrastructure. The resulting development, it needs to be stressed, would not at all resemble the new towns of the 1950–80 era: developments would be smaller-scale, but clustered along public transport corridors, and embodying locally mixed land uses for maximum sustainability.

SOLUTIONS: (3) INVESTING IN COMMUNITY INFRASTRUCTURE

The final problem is how all this is to be paid for without strain on public expenditure – and also without embarking on a fourth legislative venture that might well again prove futile, but by proceeding in a pragmatic fashion, using instruments ready to hand. We have identified three potential sources of monies for the activities of the new Regional Land Authorities: the net contributions to the Exchequer currently forecast by the Commission for the New Towns; the proceeds from Planning Obligations/Impact Fee agreements, and possible revenues from Capital Gains Tax. Together, these three sources will provide substantial, and robust, sources of additional monies for increasing the supply of housing land.

INVESTING THE NET RECEIPTS FROM THE COMMISSION FOR THE NEW TOWNS

Critical to this will be the finances of the new Regional Land Authorities. The Commission for the New Towns is forecasting net payments to the Exchequer for several more years. It would be necessary to ring-fence the present surpluses of the Commission for the New Towns so as to apply them to initial land purchases by the new RLAs. This would be equivalent to the £20 million original pump-priming grant to the Land Authority for Wales. In other words, just as with LAW, or the New Towns, the Treasury would be asked to forego an income stream for a short period in order to guarantee a very much larger income stream in future years. This is crucial to the success of the scheme, which in effect is a minimal-cost, instant-result version of the previous failed attempts to win a share of development gains for the community, and would have the additional merit of providing a modest cross-subsidy for urban regeneration.

PLANNING OBLIGATIONS

The term "planning gain" can be traced back all the way to 1909, but the current system really dates back to 1968, when the previous need for planning agreements to receive individual ministerial consent was removed, and it became possible for local authorities to negotiate planning agreements directly with landowners and/or developers. The system has survived all subsequent changes of government; Section 106 of the 1991 Planning and Compensation Act changed the language to "planning obligations" ("the Section 106 obligation forms the legal contract by which the objectives of the planning legislation are legally honoured and enforced").

Planning obligations, which came into force on 25 October 1991, come in two different forms: either a *Planning Agreement*, whereby a developer/landowner enters into an agreement with the local planning authority, or a *Unilateral Undertaking* by the landowner/developer to the LPA. Such obligations may restrict the development of the land; may require specific operations (works) or activities (uses) to be carried out on land owned by the applicant, either on the site or on neighbouring land; may require land to be used in a particular way; and may require payments to the authority periodically or at one time (there is no specific requirement within the Act that payments must relate to the land itself or to the development that is the subject of the planning application).[13]

Circular 16/91 gives guidance on what constitutes an acceptable obligation. It incorporates previous tests, particularly that the extent of what is required is fairly and reasonably related in scale and kind to the proposed development, and it states for the first time that an obligation can be used to "secure the implementation of local plan policies for a particular area or type of development" (e.g. the incorporation of affordable housing within a wider residential area). There has already been a great deal of testing in the Courts as to what is "reasonable"; the broad effect of the judgements is that the net is cast very wide.[14]

Planning gain is thus an integral part of the UK planning system, with a history stretching for nearly 30 years. It is a flexible instrument, which is both its strength and its weakness. It is more effective in periods of intense building activity than in periods of slump, but that is true of any attempt to collect betterment. It is, unusually, hypothecated: because it is regarded as a gift and not as a tax, the Treasury does not appropriate any part of it and the proceeds accrue to the benefit of the locality. Above all, it has proved acceptable to both major political parties.

EXTENDING THE PRINCIPLE: DEVELOPMENT IMPACT FEES

Planning Obligations provide a highly effective way, not of collecting "betterment" – that elusive will-o'-the-wisp that has plagued experts ever since the Uthwatt Committee report of 1942 – but simply of collecting for local authorities some of the real costs that they incur because of development. We believe that their use should be extended and codified to provide an acceptable and enduring instrument for this purpose. The way to do this has been tested over several decades across the United States: it is through so-called Development Impact Fees. These prescribe standard scales of charges to cover the costs of new development, levied on developers as a condition of permission to develop. They are normally expressed in terms of costs per dwelling; they can vary considerably, but in 1987 in the San Francisco Bay Area of California – an area of great housing pressure and high housing costs – they averaged $9110 per dwelling.[15]

In the United States, impact fees have been intensively tested through legal actions. The courts have established the principle of rational nexus: the infrastructure, to which the fees are contributing, must be relevant to the development. (In recent years, a very similar test has been applied by British courts to Planning Agreements.) The US courts have also established a whole set of supplementary criteria: need (the development must require the facilities), proportionality (fees must relate to actual costs), accountability (they must be used for the stated purpose), congruity (the fees must be used for congruous purposes, i.e. the

---

[13] Ratcliffe and Stubbs 1996, 92, 95–97.
[14] Ratcliffe and Stubbs 1996, 96–97, 101.
[15] Delafons 1990, 11.

facilities will be used by people in the development), range of costs payable (capital, not revenue), and so on. The most difficult is equity: the problem that if fees are passed on to home buyers, these buyers may pay twice over, once in fees and once in general property taxation. Often, it seems, the courts have become involved in highly arcane areas where no easy resolution is possible.

The American history suggests that the simpler the scheme, the better. There cannot be a standard formula, however; it is necessary to consider particular circumstances – the type of provision, the cost, the beneficiaries, the timing.

It seems that the system of planning agreements, which extend widely, could be revised to cover both negotiated Development Agreements and Developer Contributions. This was in fact done in the 1991 Act, except that there the contributions are called Unilateral Obligations. What would be needed, therefore, is to codify and standardise these obligations. The local authority would set out its requirement in published form, including roads, open space and school sites, as it is already encouraged to do, but it would then impose a standard scale of fees adjusted to conditions in that area. (Berkshire and Kent are trying to introduce such a system at the present time.) There is a difficult set of questions to do with what the American courts call "rational nexus", that is the strength and directness of the connection between the development and the impact, which have also been tested in the English courts (as in the case of the Tesco superstore in Witney, Oxfordshire). We believe that impact fees should be calculated on a fairly aggregate level, without too much detailed argument as to the exact relationship to the actual development; though, to introduce greater precision, the fees could be disaggregated by sub-areas, or could exclude certain categories of land (e.g. difficult brownfield land), or would apply only to developments above a certain size threshold.

There is a question that is bound to arise, as in the United States: would these fees simply be added to the price of land and thus of the housing package? This seems unlikely: in UK conditions, where land is already very expensive and a major element in the entire package, the fees would have the reverse effect, of lowering bid prices for land, so reducing the landowner's profit

There is also a question as to whether such a system would require legislation. The study by John Delafons[16] suggests that it could be done through planning guidance and/or regulation, and though a report by the Department of Town and Regional Planning at the University of Sheffield, published in 1994, suggests that it would require modifications to the existing statutory framework, we accept Delafons' conclusion (which however would need to be tested). The DETR would produce guidance containing model national clauses, with suggested provisions for local variation. A scale of local charges would then be adopted, following publication and provision for objection, and probably covering entire regions; this might be subject to call-in by the Secretary of State. These changes might usefully be wrapped up in a standard payment accompanying the grant of planning permission.

---

[16] Delafons 1990.

Such a standardised system might actually be welcome to the development industry, since it would provide a degree of certainty. There are a number of questions that would need further consideration: the precise method of calculation (unit, area, floorspace?), the precise relation to the benefit that the new infrastructure will bring, and the revenue that may flow from its use; the difficult problem of the application to sites of marginal productivity; and the question of how the whole scheme would be regulated. Further, care would need to be taken that such a scheme, nationally collated, did not in effect constitute a form of taxation, thus raising the traditional Treasury objection to hypothecation. The way to resolve this might be to specify that, as now, it would always be open to putative developers to opt to reach negotiated agreement with the planning authority as an alternative to a standard fee payment.

Such a scheme, as well as being a rough and ready solution to the problem of charging for public infrastructure provision, has an additional merit: it could make local authorities more enthusiastic (or perhaps, more accurately, less unenthusiastic) about giving planning permission for development. In the more pressured regions, NIMBY political pressures are allied with fears about the implications of development for local public spending. This proposal would not remove that concern. But it could help ameliorate it.

However, if local district authorities were to collect all the income from Impact Fees, as they do now from Section 106 obligations, that would create a severe inequity: in those areas of the country where provision of infrastructure would be especially difficult and expensive (such as contaminated sites), the Impact Fees could be so high as to inhibit development. Therefore there is a case for abstracting some small part of the gain for the use of the Regional Land Agencies, serving as a redistributive device to these difficult areas.

CAPITAL GAINS TAX

Impact Fees would be levied only to cover local authorities' specific obligations, and would represent a codification of present arrangements for collecting Planning Obligations. But, on the basis of American experience, they would cover no more than a portion of the actual infrastructure costs imposed on local authorities by developments, and in no sense would they cover the wider problem of unearned increment in land values that follows development permissions. To deal with this, we would suggest that the least complex way would be to impose a special level of Capital Gains Tax in respect of gains in land values, as was actually done during the 1970s before the introduction of Development Land Tax. Such a tax, at say 50 per cent, would be net of any contribution the landowner or developer might make by way of Impact Fee, Unilateral Undertaking or Planning Agreement. Logically, then, like the Land Authority for Wales before abolition of the development Land Tax, English Land would buy land net of this tax.

There are, however, limitations to the use of CGT. It is not applicable to pure land dealing companies, and so would not catch much of the activity that takes place. To capture this would need a return to a solution similar to those that were adopted in the 1960s and 1970s. For reasons already stated, we do not believe that it would be politically wise to go again down that road. Therefore, our preferred solution embodies three elements:

1. A stress on major schemes, both for redevelopment and new development, where unambiguously the values created by the scheme would be disregarded in the price paid for land.[17]
2. For smaller developments, use of codified impact fees.
3. For very small developments, where there was no material impact, the owner would keep the development value.

Thus, effectively, we are recommending an extension and codification of the existing system. The novelty would lie in the greater emphasis on major schemes, not of the traditional new town type, but constituting sub-regional development and redevelopment clusters.

THE NEW TOWNS MECHANISM

Finally, there is the new towns mechanism. In the PARDs – wide but often discontinuous zones or corridors, generally along major public transport routes, which are especially suitable on grounds of sustainable urbanism for concentrated packages of regeneration and new development – the Pointe Gourde principle will operate: the New Towns legislation will apply, and the Regional Land Authority will not pay any value that is ascribable to the existence of the scheme. Generally, therefore, this land will be subject to compulsory purchase. In addition, of course, sales of lands to developers will include standard Impact Fees, which will generally be uniform within each PARD but may therefore represent an average, allowing an element of cross-subsidy for the treatment of more difficult sites.

---

[17] This does not represent a 100 per cent tax on development gain, since the courts have established that it is necessary to take into account what would have been the most likely permitted use in the absence of the scheme.

CHAPTER 11

DO-IT-YOURSELF NEW TOWNS

It is usually landowners and speculative developers who attempt to subvert the planning machinery for their own benefit. But both the present authors plead guilty to the charge of trying to expand pinholes in the blanket coverage of land use legislation, in the interests, not of property speculation, but of the variety of people whose chosen lifestyles – or whose involuntary poverty – fail to fit the assumptions of the planners.

Most professional activities have a primary duty to individuals, but professional planners are almost exclusively and obviously the servants of authority and the state, national or local. It is a characteristic of representatives of the state to seek to extend their powers of control. An obvious example in the planning field is the expansion of official decision-making from land use to aesthetics. Yet the founding father, Ebenezer Howard, as F.J. Osborn explained to Lewis Mumford in March 1969, "had no belief in 'the State', and though he had a belief in the essential goodness of human nature, he didn't expect that any environmental change would turn us all into angels".[1] In the same month, a quartet of iconoclastic authors in the weekly *New Society* launched a squib called "Non-Plan: An Experiment in Freedom", which urged that there should be "a precise and carefully controlled experiment in non-planning" in which "people should be allowed to build what they like". It concluded that "except for a few conservation areas which we wish to preserve as living museums, physical planners have no right to set their value judgement, against yours, or indeed anyone else's. If the Non-Plan experiment works really well, people should be allowed to build what they like."[2]

This article was seen as a merry prank, and was ignored. But in 1975, given the opportunity to address the Garden City/New Towns Forum, the chance was seized to link this aspiration to the actual experience of the prewar plotlands in Britain and to the self-built unofficial cities of Latin America, Asia and Africa.[3] This paper claimed that there were already "large numbers of people interested in alternative ways of making a living: looking for labour-intensive low-capital industries, because capital-intensive industries have failed to provide them with an income", and it argued that "One of the essentials of a do-it-yourself New Town would be a relaxation of building regulations to make it possible for people to experiment in

---

[1] Hughes 1971, 453.
[2] Banham et al. 1969, 443.
[3] Ward 1990, 15–35.

alternative ways of building and servicing houses, and in permitting a dwelling to be occupied in a most rudimentary condition for gradual completion."[4]

This plea might have met the same deafening silence as the advocacy of plan-free zones, but it won some support (and plenty of opposition) among the staff of Milton Keynes Development Corporation. By chance that body's next government-appointed chairman was also the chairman of the Town and Country Planning Association. This was Lord Campbell of Eskan, an industrialist who was also a Labour peer. At the TCPA annual general meeting in 1978 he recalled that he became a socialist in the 1930s, when he became a director of a company operating the sugar industry in the then colony of British Guiana and was horrified by the "wholly morally, socially and politically unacceptable" conditions in which his firm's employees lived.

> But with the price of sugar and the profitability of the company as low as it then was, by no stretch of the imagination could we afford to build proper houses for everybody. And then one day an idea struck me: why not lay out building plots on spare land adjacent to each estate – about ten to the acre; put in roads, drainage and water standpipes and let each family have a building plot, the materials of their present abode – all the buildings were in wood – some free paint, a present of £25 and an interest free loan of £250 to be repaid slowly out of wages. These modest figures went up later on. The scheme took on like wildfire, and within a few years virtually everybody was re-housed in the new areas. As there was no effective planning or building regulations in the Colony, every sort of house under the sun was built, from corrugated iron shacks with the rest of the space on the plot used for cattle and goats, to palatial houses costing £10,000 or more.[5]

Learning from this experience, and commanding the "astonishingly fresh and relevant" principles of Ebenezer Howard's vocabulary, Campbell urged that the TCPA should campaign for a small country town in its own belt of market land, as a Third Garden City, since "such a project could take on to the next century the ideas that the TCPA gave to this one". He suggested that an approach should be made to his Development Corporation for the provision of two grid squares (about 500 acres or 202 ha) of its undeveloped land for such an experiment. The idea was presented in a crowded marquee at the Comtek Festival at Milton Keynes and 60 prospective residents held a further meeting on the same day to form the Greentown Group. Meanwhile the TCPA set up a series of working parties to consider nine aspects of such a settlement: housing, employment, farming and landscape, personal services, utilities, communications, community structure, finance and development, as well as the implications for inner cities.

The Development Corporation identified a site of 34 acres (13.7 ha). The TCPA thought this too small, "and the Corporation was asked to agree to the release of adjoining land as well in due course. In response the Corporation, sensitive to local opinion, said it was "most unlikely" to do so.[6] The TCPA, faced with an ultimatum, withdrew from the venture in 1981. And in the climate of inflated land prices,

---

[4] Ward 1990, 32.
[5] Campbell 1978, 2.
[6] Hardy 1991b, 180.

As a last-ditch attempt to salvage something from the experience, those who were left in the Greentown Group decided to reduce their claim to that of a 6-acre site for a self-build residential scheme. Abandonment of the mixed-use proposal, with its self-reliant element, served to appease the Borough Council . . . Negotiations were finally brought to an end . . . on the grounds that the Group had failed to present a convincing financial management strategy.[7]

The attempt to insert a do-it-yourself Garden City into Milton Keynes had several lessons. It was in those days a matter of principle for the government-appointed Development Corporation to defer to the elected local authority and avoid confrontations. The Council was alarmed by Greentown. Would they inherit a shanty-town? Would they become hosts to a hippy commune? Don Ritson of the Development Corporation explained that "We can't get planning permission, even in outline, without a clear statement of what is to happen on the site, but if we specify what is to happen we are limiting in advance the aspirations of the people who we expect to settle there. And the whole idea is to give *them* the freedom of choice."[8]

In any case, the mood of the Development Corporation itself was obliged to change as the incoming government of 1979 set about "turning the warm-hearted, motherly, public-service-oriented Milton Keynes of the 1970s into a slim-jim, self-financing, property investment machine designed to suit the commercial disciples of the 1980s".[9] The concept of the Development Corporation as a privileged promoter of experiment had evaporated.

But in another new town hope rose again in 1980 when Lord Northfield, the chairman of Telford Development Corporation in Shropshire, wrote to David Hall, Director of the TCPA, raising the possibility of providing land for an alternative community on a Telford site ravaged by two centuries of old coal workings and considered unsuitable for the ordinary process of development. The corporation's chief planner joined one of the working parties of the TCPA's new communities project. The Development Corporation set aside 250 acres (100 ha) of third-grade land, for which no other potential use had been found. Gillian Darley, reporting later to the Rowntree Foundation, was to explain that all such ventures depend upon the accident of a landowner who is prepared to defer the financial return on investment, "or who is able and willing to hand over the land at a bargain price as a gesture of goodwill in the venture, as the Telford Development Corporation originally did for the prospective Lightmoor community there, where the winding up of the Corporation has meant that the offer for 2–50 acres has been reduced to 23 acres and a further, equally small, second phase".[10]

The experience of the pioneering 14 families on their 23 acres (9.3 ha) illustrates the wisdom of the establishment from the start of the ideology of Non-Plan or of the legislation-free "Do-It-Yourself New Town" (Figures 50 and 51), for specialist professional advice and the exploitation of special clauses in the legislation had to be sought out just to get Lightmoor out of the ground:

---

[7] Wood 1988, 24.
[8] Ward 1993, 129.
[9] Bendixson and Platt 1992, 181.
[10] Darley 1991, 13.

Figure 50 *Lightmoor. Building the famous community in Telford New Town, built on reclaimed land by the residents; achieved on TCPA initiative after a failure at Milton Keynes.* Source: Town and Country Planning Association.

Figure 51 *Lightmoor. The completed development; the fences keep in the chickens and goats.* Source: Town and Country Planning Association.

Telford Development Corporation obtained a Section 7.1 approval under the New Towns Act from the Department of the Environment for the establishment of a new community at Lightmoor, which permits mixed use of the site. Each individual plan for house and/or workshop has to be approved under Section 3.2 of the Act . . . The crucial element that distinguishes this procedure from normal planning control is that the 7.1 approval allows residents the opportunity to set up their own home-based enterprises either in a workshop sited on their plot or incorporated in their home designs. This, coupled with the half-acre plot size, on which there is room to keep livestock, which could provide the family with eggs, milk, cheese, is meant to enable residents to opt, if they so wished, for a belt-and-braces economy.[11]

Readers will be astonished that these modest, worthy aspirations should be faced with such difficulties. In fact the biggest obstacle, both in the unrealised Greentown project at Milton Keynes and at Lightmoor, was the fear they aroused in the local authorities. And the most grotesque irony at Lightmoor was the fact that the activities of the 14 pioneering families on this derelict site (which aroused huge interest nationally, so that members had to announce that casual visitors would only be welcome on the first Sunday afternoon of each month), had so upgraded the notional value of the future extension, that in the new market-oriented climate of the mid-1980s, it was considered as too valuable for such a marginal settlement of low-income people.

This left the pioneers who had taken the best advice and evolved a "legal framework that has caused much frustration and expense" and an elaborate company structure intended for the management of 400 houses, not the 14 that emerged. But this tiny project aroused immense interest precisely because it represented the fulfilment of such widespread dreams. Locally it was called The Good Life Village, half derisively, half enviously. It was anticipated that there would be an explosion of emulations and imitations among people seeking alternative ways of living and of housing themselves.

It did not happen. Not because the aspirations were not there, but because a series of factors ensured that nothing could happen. One of these is the price of land, artificially enhanced by planning policy. (In the South East of England, site costs amount to 65 per cent of the cost of a new house.) Another is the NIMBY factor: we are hostile to the idea of neighbours with different aspirations. But the implacably determining factor is planning policy. The most persuasive challenge to this situation comes from Simon Fairlie, an advocate of "low-impact" rural development.

Fairlie is one of those people anxious to experiment with "permaculture" as a land use system, feeding themselves organically on a small patch of ground. The vocabulary is different, but the aspiration closely resembles that of the simple-lifers of a century ago described in Chapter 5. Simon Fairlie was one of a group of friends who rented a house with a large garden on a country estate, but were evicted to make room for a golf course. After living in a van for two years, he joined another group and bought a smallholding with no house attached. They pitched

---

[11] Broome and Richardson 1991, 93.

seven tents and set about cultivation. The result, he reported, was that "In the two years since we moved onto our land, we have been through almost the entire gamut of planning procedure: committee decision, enforcement order, stop notice, Article 4 application, Section 106 agreement, appeal, call in by the Secretary of State and statutory review in the High Court." But his purpose is not to demonise the planning machine. He believes in it because he knows that without it speculative developers would have completed the destruction of the countryside initiated by the farmers, subsidised to destroy woodlands, wetlands, hedges and wildlife. He argues that,

> If permission to build or live in the countryside were to be allocated, not just to those who can afford artificially inflated land prices, but to anyone who could demonstrate a willingness and an ability to contribute to a thriving local environment and economy, then a very different kind of rural society would emerge. Low impact development is a social contract, whereby people are given the opportunity to live in the country in return for providing environmental benefits. Planners will recognise this as a form of what they call "planning gain". The mechanisms to strike such a bargain are for the most part already written into the English planning system and there is thus no need for any major structural changes.[12]

He is optimistic, of course, but he seeks nothing more radical than a simple reinterpretation of existing policies. He argues, for example, that Simplified Planning Zones (SPZs), introduced in the legislation by the Housing and Planning Act of 1986, as perhaps a belated tribute to the proposals recalled at the beginning of this chapter, could perfectly well be used to manage sites for this kind of low-impact development, whether for residential or mixed use, with two main advantages:

> Firstly, the occupants of the SPZ are free to build anything, anywhere, within the constraints laid down by the schemes thereby combining maximum scope for ingenuity with minimal interference and paperwork. And secondly, the development is a "one-off", a clear exception to the planning regime in the rest of the district, and therefore cannot be used as a precedent.[13]

Fairlie's aim, as he says, is not to dismantle the present planning system but to exploit loopholes like Simplified Planning Zones – introduced for other purposes and described in the Department of the Environment Planning Policy Guidance note (PPG 5) – to enable local authorities to foster experiments in low-impact rural development, "Some of them carried out at the margins of society, others designed to cater for more conventional people."[14]

The point is well made. We can be certain that the early decades of the next century will see an increase in the number of people aspiring to a Green lifestyle and to house themselves in what they perceive to be a low-impact fashion which they will claim is far more sustainable than current interpretations of

---

[12]   Fairlie 1996, 68.
[13]   Fairlie 1996, 124.
[14]   Fairlie 1996, 130.

Figure 52 *Single-Person Dwelling. An experimental house, designed by Pat Borer using the approach developed by Walter Segal, and built by students of self-build housing in the Centre for Alternative Technology at Machynlleth in Wales.* Source: Brian Richardson.

sustainability. There will also be New Age travellers anxious to grow roots, and tired environmental protesters anxious to retire, having converted public authorities to their point of view, all seeking to house themselves in a rudimentary way. Their faces will not fit in the offices of the mortgage-lenders, but, in the new insecurity of the flexible labour force, nor will those of far more conventional people.

The expansion of self-build housing among people with time on their hands but very low incomes has been one of the few cheerful aspects of the late twentieth-century housing scene in Britain.[15] Like the failed attempts to get a Third Garden City out of the ground, first at Milton Keynes and then at Telford, the most modest of self-build projects have to battle against administrative assumptions not designed to meet their needs. Yet the promoters of an effort to enable unemployed young people to house themselves, the North Tyneside Youth Self Build Enterprise, make such remarks as "It was the most worth-while enterprise I have been involved in during the whole of my professional life." And they will often add, "You should see how it transformed the lives of the people involved."[16]

The architect Walter Segal became famous late in life through his efforts to devise a cheap, quick and simple approach to timber-framed house-building that anyone could use. When he finally persuaded a local authority to offer a site to a

[15]   Broome and Richardson 1991.
[16]   Ward 1993, 131.

mixed bunch of people from its housing waiting list, the result, after endless bureaucratic delays, was a triumph, immensely relevant in a country with both a history of housing disasters and with many people in unsought idleness. Instead of being dismayed, as many architects would be, at the "countless small variations and innovations and additions" to the designs he had worked out with each individual family, he rejoiced that, as he put it's "there is among the people that live in this country such a wealth of talent," and he found it unbelievable that this creativity would continue to be denied outlets[17] (Figure 52).

Most of us would share his mystification at the way in which the assumptions of the governmental machine, as well as those of the land market, continue to ignore the role that could be played in the sustainable Social Cities of tomorrow by the people, so close to the world of Ebenezer Howard, who have aspirations of their own.

---

[17] Ward 1990, 80.

CHAPTER 12

NOT COUNTING NIMBIES

The late Professor Leopold Kohr, in his celebrated book *The Breakdown of Nations*, advocating city regions as the twenty-first century equivalent of the nineteenth-century cult of the nation state, included a chapter with the title "But will it be done?" Its text consisted of the single word "No!".[1] In precisely the same spirit we could ruefully conclude that tomorrow's version of Howard's Social Cities, catering for the projections of household formation, *can* be provided through the intelligent exploitation of railway and light rail networks, but is unattainable, simply because of the growing strength of the anti-development lobby. In Chapter 10 we suggested that it will be important to ensure that the process of developing regional housing strategies is not compromised by local NIMBY pressures, which are all too likely to be present. In fact, the "pull-up-the-ladder" group of recent arrivals, as Peter Ambrose labels them, are one part of a very diverse conservation lobby, and, as he explains:

> The group has come to these rural areas primarily to enjoy leafy seclusion. The last thing they want is another group of arrivals. In other words they are rigidly opposed to any more housebuilding if it will spoil their view or possibly have an adverse effect on property values. They may well be in favour of more development in the general vicinity, perhaps a motorway giving better access to centres of employment, but they will often use their considerable expertise to organise resistance to development in, or within sight of their particular village.[2]

Other components of this lobby include what he calls "the genuine core of conservationists, those interested in the rural heritage, the beauty of the landscape, the wildlife, the preservation of hedges and so on", as well as another segment of the conservation coalition, those people who "have an interest in preserving the old hierarchy of social relations based on position in the agri-cultural system of production and the various skills and crafts necessary to support it". Several influential interest groups speak for this urge to preserve the "country way of life". One is the Country Landowners' Association, whose name explains its interest, and another is the Council for the Protection of Rural England, whose employed research staff often have a broader and more comprehensive approach to rural land use than the local activists.

---

[1] Kohr 1957, 197.
[2] Ambrose 1992, 186.

There is some evidence that the very word "countryside" is enough to win support. On 10 July 1997, a newly formed body called the Countryside Alliance persuaded 250 000 people to rally in Hyde Park to defend "rural values" from ignorant town-dwellers, and to hear the Leader of the Opposition declare that a private member's bill called the Wild Mammals (Hunting and Dogs) Bill "is a divisive measure creating two nations by setting town against country". According to *The Guardian* on the following day, the biggest applause on that occasion went to the cry of farmer Willie Poole, "Stop letting these townie buggers grind us down".

THE TRUE DESPOILER: BIG FARMING

In real life it is the farming industry, and by no means the ignorant town-dwellers, that is responsible for the despoliation of the countryside in the decades since World War II. It was a farmer and Conservative Member of Parliament, Sir Richard Body, who reminded the Town and Country Planning Summer School at Lancaster in 1993 that "the intensification of agriculture in the last 25 years has gone ahead faster and more furiously in the United Kingdom than in any other member state of the EC", and he read out to the assembled planners what he called "the woeful litany of statistics of the damage inflicted on the rural environment by government subsidies to farmers". These included:

- 130 000 miles (210 000 km) of hedgerows ripped up
- 40 per cent of our ancient woodlands gone
- seven million acres (2.8 million ha) of pasture-land ploughed up
- over 95 per cent of our wetlands drained
- 875 miles (1410 km) of stone wall destroyed
- 95 per cent of the downlands of southern England gone
- 180 000 acres (73 000 ha) of moorland ploughed up

"Some of us," he said, "have made such an uproar about this agri-vandalism that in recent years we have seen the introduction of several schemes to undo the damage." It infuriates participant observers like him that, having subsidised the owners of rural land to do all this damage in the name of increased food output, we are now "paying the farmer to manage the countryside and thus protect the rural environment".[3]

Now, as we have discovered in Chapter 7 of this book, on the basis of official statistics, the quantity of land set aside in 1995 under the EU's agricultural policy, and deliberately excluded from producing food, or anything else with an economic value, is three times the amount of land needed to accommodate all urban development predicted for the coming quarter century.

---

[3]   Body 1993, 62–66.

Yet the powerful countryside lobby is silent about the heavily subsidised agricultural industry and has ensured that migration into rural England is confined to the affluent; thus, while rural life has been transformed, as we have commented earlier, the bottom 25 per cent of the rural population are just as much left out as they always were. Preserving the precious rural habitat means protecting the affluent who can afford rural house prices, but do not use the declining public transport system nor patronise the surviving village shop, and whose children do not attend the threatened village school.

In the week when the organised defenders of "rural values" assembled in Hyde Park, the Rural Development Commission issued a report on the situation of young people in ordinary low-income rural families who have to remain in the parental home well into adult life, as the "countryside" has no place for them and the only alternative is to rent rooms in the nearest town.

The report's concluding sentences were:

> Young people in rural areas, and the projects assisting them, are confronted by social, transport, housing and labour market structures that they themselves cannot change. Addressing these structural issues would address the roots of the problem faced by young people in rural areas with respect to housing.[4]

These deprivations are precisely those listed a century ago by Ebenezer Howard in his Three Magnets diagram, and precisely those which would be alleviated by new Social Cities clustered on rail routes. Yet such developments are among the likeliest to be most energetically opposed by the "defenders" of the countryside.

THE NEWEST PRESSURE GROUP OF ALL

But the most recent rural pressure group is the most challenging of all. In the 1960s the government of the day commissioned an enquiry into public participation in planning. The Skeffington Committee, in its report, defined the purposes of planning as "to set the framework within which houses, roads and community services can be provided at the right time and in the right place".[5] Public participation has been hard to elicit, except in the defence of existing vested interests, but a new form has emerged dramatically in the 1990s.

This is in the form of an environmental protest organisation, an intriguing coalition of old and new defenders of rural habitats, with no automatic concern with either property values or self-interest. Skeffington failed to anticipate the determination of the road protesters. But, in the last decade of the century that had seen the triumph of the private motor vehicle, a militant counter-attack has arisen in the form of the road protesters.

---

[4]   Rural Development Commission 1998.
[5]   GB Department of the Environment 1969.

They are young, dedicated, witty, and very resourceful. And their strategies, of building tree-houses in threatened trees or tunnelling below the route of the earth-movers, captured the public imagination. There were prolonged struggles over a series of road-building contracts and the popular heroes were young protesters known as Animal and Swampy. Two days before Swampy was driven out of his Fairmile tunnel, the *Daily Mirror* (29 January 1997) carried a feature celebrating "The Battle to Stop Cars Taking Over Britain", and an editorial explaining "Why Animal is Right," arguing that "Britain faces a choked, gridlocked future". And, as though to stress the point, Steven Norris, a former Conservative transport minister (who was not seeking re-election) admitted in a *Panorama* TV programme in March 1997 that in his view the protesters in tree-houses along the route of the Newbury bypass "were right" and that the road should never have been built on the chosen route. He said, "I think it's fair to say that the formula was more motorist-based than it should have been and that it didn't apply the same kind of cash values to environmental considerations which it did to motorists' inconvenience."[6]

With a change of government some road proposals were cut and some were shelved, and the appropriate rhetoric was used to recommend a reduction of car-dependency and a greater reliance on public transport. However, there was no indication of any commitment to a policy of winning people back to public transport and of planning future settlements around potential rail and light rail networks.

This demands something very like Ebenezer Howard's concept of a century ago. Stephen Potter of the Open University New Towns Study Unit found that, while both Garden Cities and New Towns abandoned Howard's transport priorities,

> True mobility is dependent upon unhindered access and the ability to use all the major forms of transportation that are available. None is so minor as to be unworthy of consideration in the planning context, or anticipated to be in the future. But the operating requirements of these transportation systems are such that severe design conflicts arise in the urban form that is optimal for each. From our experience of building new town type projects since Ebenezer Howard began the New Towns movement at the turn of the century, it is clear that only one set of priorities can succeed in *fully* resolving these design conflicts. That is to give the pedestrian and cyclist primary consideration, followed by that of public transport with the flexible and adaptable car fitting into the thus determined structure. This was the set of priorities that Ebenezer Howard proposed in 1898, which was rejected by those who built the Garden Cities, and to which we have only just returned and have confessed to be a valid approach. In terms of the principles and philosophy of transportation planning for new communities, far from being at the zenith of 80 years' work (as many might have us believe), we have only just stumbled back to the beginning![7]

The big task is going to be that of winning the road protesters to this view. In the early days of the Garden City Association, it embarked on a hectic campaign of

---

[6] Anon. 1997a, 44.
[7] Potter 1976, 67.

propaganda to win citizens around. It is recorded that "From 17 August 1902 to 17 May 1903 the Association offered over 260 meetings or lectures illustrated by 'limelight views' in locations ranging from Cheapside to Edinburgh."[8] The experience of the advocates of Do-it-Yourself New Towns, described in Chapter 11, suggests that, a century after Howard, we lack that proselytising energy, and the appropriate style of propaganda to win over, not so much the dedicated road protesters but their innumerable sympathisers, who recognise that new settlements are needed but are bewildered about the form and locations that they should adopt.

The planning system automatically favours the established speculative developer who automatically assumes the white-collar male home-owner in secure employment. But the latter is a diminishing species, not only demographically, but in terms of the flexible labour market sought by government for two decades. Developers have sought to earn their quota of providing "planning gain" through including in their proposals a few sites for "affordable housing" to be provided by housing associations for rent. The very concept of affordable housing is another reminder of the fact that, as stressed throughout this book, we have failed to come to terms with the vital issue of the development value of land.

SQUARING THE CIRCLE

How do we square this particular circle? Are there ways of resolving the interests of the NIMBIES and of all the people – between two and three million of them, on any reasonable count – who are almost certainly clamouring to join them? Surprising as it may seem, perhaps there are.

The fact is that it has been done, once before. In the 1930s the most distinguished British planner of his day, Patrick Abercrombie, chaired the Council for the Preservation of Rural England (as it was then known) and was a member of the Council of the Town and Country Planning Association. More than that: with other like-minded people, such as F.J. Osborn, he helped shape a coalition of the countryside preservation lobby and the new town campaigners. Osborn at one point thought that Abercrombie had sold the pass, by calling in the County of London Plan for densities that were too high to house families with children; but then he forgave him for the Greater London Plan, with its scheme to move more than one million Londoners to new and expanded country towns. It was this alliance that produced the historic postwar consensus, balancing protection of the countryside on the one hand, with a programme for nearly 30 new towns on the other.

The background to this history of coalition-building is significant. The CPRE and the TCPA were no less committed to their respective causes in 1938 than they are in 1998. Osborn, in particular, was not by nature a compromiser. But

---

[8] Moss-Eccardt 1973, 28.

they faced a common enemy in suburban sprawl, above all around London, and the pepperpotting of houses over the countryside outside the capital (including, it must be admitted, the "plotlands"). Indeed, everyone who was anyone (meaning anyone who wrote letters to *The Times*) opposed those kinds of developments, and saw them as a threat which called for a common front. Osborn, who in 1939 like everyone else talked in military parlance, actually called it the Planning Front.[9]

The resulting common platform essentially had two elements: on the one hand, defend most of the rural acres by effective countryside planning which would in effect bar new building; on the other, develop compact new towns and town expansions to provide for the overspill from the cities. It was this historic alliance that produced the New Towns Act of 1946 and then the Town and Country Planning Act of 1947, which provided the draconian powers (above all, through its solution of the so-called Compensation and Betterment problem[10]) to establish green belts and other restrictions, thus effectively stopping further suburban sprawl on the 1930s pattern.

We need to ask, what is different now? What has happened in the intervening half-century, that might inhibit or prevent another such alliance of the principal campaigning organisations concerned with the use of our land? One answer is that we are four times as affluent, so that the numbers of people wanting to invade the countryside are much greater now than then. But this is an over-simplification: most people may have been less affluent in the 1930s, but they found it far easier to afford new housing, which was cheaper in relation to income than ever before or after.

It is more likely that, after this half-century of effective planning at county level, most country-dwellers simply feel more secure and less inclined to compromise. But the essential dilemma they face is no different now than then: if they fail to take action, they may well face a worst-case outcome, in which constant drip-feed pressure for development causes the outcome they most fear: tens of thousands of penny-packet developments on ill-conceived sites, won on appeal after long and grinding battles. Indeed, this is not just a possible outcome; it is almost certain, if the authorities fail to take bold and resolute action to provide building land on the scale that is needed.

Again, there are parallels from the past, described in Chapter 3. In the late 1940s, Hertfordshire agreed to take four of the eight new towns proposed in the Abercrombie Plan, though many of its residents were bitterly opposed to them – as

---

[9] Hardy 1991a, 215.

[10] Compensation referred to the payments that central or local government must make for compulsory purchase of land, for a new town for instance. Betterment referred to the increase in value that arose through public works like a new road or railway, or even through planning restrictions like a next-door green belt, which accrued to landowners through no effort of their own. The 1947 Act effectively nationalised development values, providing for once-for-all payoffs to alleviate hardship to landowners; it thus allowed land to be bought at existing agricultural values, and provided that increases in value should pass in their entirety to the community through a Development Charge.

Figure 53 *"Silkingrad". The defaced train station signs at Stevenage in 1947, shortly after its designation as the first postwar New Town; NIMBYism was already alive and well, though no one had then heard the expression.* Source: Town and Country Planning Association.

the early years of Stevenage, with the violent demonstrations and the Silkingrad incident, so well demonstrate (Figure 53). Twenty years after that, in the mid-1960s, Buckinghamshire effectively agreed to accommodate the new city of Milton Keynes as part of a package that included preserving the southern half of the county against large-scale development. The pressures were at least as ferocious in the late 1940s and the mid-1960s as they are in the late 1990s, but far-seeing public authorities planned wisely with the interests of their own people in mind. What they could do then, so now can we.

Except that now, perhaps, the counties are no longer sufficient units: they are too small; in particular, since the Gummer reorganisation of 1993–97, they are too divided or punctured by separate urban planning authorities; and their boundaries no longer correspond, in many cases, to the areas for which effective sub-regional planning needs to be performed. Regions would be better, and that is why the government's 1997 proposals for directly elected regional development authorities have been received with such enthusiasm. But they are unlikely to be in place before 2002. And, in the South East, it is not certain that even regional authorities will be sufficient: in key areas, like Milton Keynes–Northampton–Wellingborough, the need is for sub-regional authorities that will straddle the boundaries of the current standard regions.

So the aim now should be to re-create that Planning Front, or Planning Alliance, to meet the challenge of the new century.

Self-Sustaining Coalitions

Such a coalition might be called conservative in the strict sense: it will exist to guard against a worst-case outcome for most of its members. But there is also a case for a more radical kind of coalition, call it a left-wing coalition if you like, with an agenda which might alarm some of the above people but might very well appeal to many others. For, if we are ever to win popular support for the programme set out in the last chapter, we need as a matter of both principle and strategy to include the aspirations of the excluded minority. It makes sense to provide for self-builders, subsistence growers and permaculture advocates in tomorrow's sustainable settlements just as it makes sense to anticipate the growth in Britain of the trend to be found all over the United States: Farmers' Markets.

The slogans of such a coalition would be self-help, self-build, self-sufficiency, and self-sustainability. The central theme would be autonomy: the construction of settlements that are, as far as feasible at the end of the twentieth century, self-governing and self-contained.

These settlements, it hardly needs saying, would not be new plotlands. The politics of the 1990s would deny that possibility: they would specify that – like the earlier failed experiment at Milton Keynes and the later successful experiment at Telford – they occupy sites defined by the planning system. The most likely sites, as those earlier experiments show, are neighbourhoods of new towns or settlements planned and built under new towns legislation; the development corporations find it easier to do this than the average district council, looking nervously over its collective shoulder at an anxious electorate. The building regulations of the 1990s, with their stringent demands in terms of public health and insulation, would apply to the houses to be built here, as they would to all others. But these houses would share with the plotlands one key feature: that people would themselves build the houses they wanted to build.

The way to do that would be to copy the formula that built the plotlands, but to adapt it to the conditions of the late 1990s. Young architects would be encouraged (in fact, they would need little encouragement) to produce model designs which could be bought off the shelf, like the pattern books of the 1920s. But, through annual competitions with prizes and lots of publicity, they would be encouraged to achieve the highest levels of design. Their designs would be specifically aimed at self-build, with generous use of DIY components bought from the Homebases and B&Qs and Do-It-Alls, like roof trusses and window frames. Selected areas, easily accessible to public transport, would be made available with self-build plots, ready-serviced, and with an offer of automatic planning permission for approved designs.

There are other features that could and should be incorporated in at least some of these developments. The most important is found in many early Garden Cities and Garden Suburbs of the 1905–40 vintage: at Brentham Garden Suburb in Ealing, at Plessy-Robinson outside Paris, and at Römerstadt outside Frankfurt.[11] This is an allotment garden, which ideally would be provided in the communal

---

[11] Ealing Garden Tenants 1912; Fehl 1983; Hall 1988.

open space in the middle of a superblock, entirely surrounded by houses and their own small private gardens. It would answer the insistent call for organic food from an increasingly sophisticated and worried public.

THE RETURN TO ALTERNATIVE AGRICULTURE

But the principle could go further, to follow the original prescription in Howard's *To-morrow!*, and found in reality at Brentham and at Römerstadt: more generous smallholdings, in effect small farms, would be provided in adjacent agricultural land forming part of a green belt around each separate settlement. Here local people would grow organic produce for sale, thus generating additional income, above all income to supplement low basic wages. This produce would be sold in a farmer's market, which would form an integral part of each settlement's town centre, perhaps forming a specialised extension to a local superstore.

This is not a nostalgic fantasy, derived from some vision of the remote Middle Ages. On the contrary: it is an idea whose time has very much come. Joan Thirsk's magisterial study of the history of alternative agriculture shows us that the idea has taken root four times in our history: after the Black Death; in the century after 1650, which inspired Gerrard Winstanley and his Diggers; in the great agricultural depression which began in 1879, and which so influenced the thought of Howard and his original followers; and since the 1980s, as a reaction to the unparalleled environmental destruction wrought by industrialised agriculture. Thirsk finds that, in each period, alternative agriculture has improved our diet and has been a key source of technical innovation. And the movement has hardly if ever been top-down. She concludes:

> judging by the experience of the three previous phases of alternative agriculture, the strong assumption of our age that omniscient governments will lead the way out of economic problems will not, in practice, serve. The solutions are more likely to come from below, from the initiatives of individuals, singly or in groups, groping their way, after many trials and errors, towards fresh undertakings. They will follow their own hunches, ideals, inspirations, and obsessions, and along the way some will even be dismissed as harmless lunatics.[12]

Now, she believes, their time has come again. For there is a major sea-change in the tastes and habits of the British population, arising from widespread affluence and the desire for a healthy lifestyle. More and more people are experimenting with a wider and wider variety of foods, partly because they travel more, partly because the media find that food articles sell papers and boost ratings; more of these people take an intelligent interest in the effects of food on their health. People's tastes are becoming more subtle with affluence; witness, for instance, the explosion of different kinds of bread on the bakers' shelves.[13]

---

[12]   Thirsk 1997, 256.
[13]   Thirsk 1997, 260.

So there is an increasing role for innovative and intelligent farmers, even obsessive ones, who want to produce new kinds of food and to produce all kinds of food in more wholesome ways. And these are not the big farm barons. Thirsk, an economic historian with unique knowledge of British agriculture over seven hundred years, is well aware of the irony of the situation. For this happened a century before; and the visionary who foresaw it was a contemporary of Howard's, who published his own vision the same year as Howard, and who shared his basic beliefs:

> In the late nineteenth century phase of alternative agriculture, Peter Kropotkin argued most eloquently in favour of labour-intensive work on the land. Demanding more horticulture, he stressed first and foremost the common sense of growing fruit and vegetables at home to replace rising imports, but he also pleaded the good sense of providing work for all . . . The same may be said today. A notable characteristic of many horticultural ventures is again their labour-intensity, and in a climate of opinion which also acknowledges labour as a therapy, it is striking how often the horticulturists themselves stress the satisfaction of their work, despite the hard manual labour.
>
> Since far-sighted individuals have forecast the impossibility of restoring full employment now that modern technology is daily reducing the work required, we plainly await another Peter Kropotkin to pronounce the same lesson all over again. The continuing obsessive drive to foster technology and shed labour at all costs belongs appropriately to the phase of mainstream agriculture, and not to the alternative phase.[14]

The alternative agriculture movement is now as strong as it ever has been in the past 700 years, but the revolution in our mind-set and in our habits – marked as it may seem – is as yet in early days; it will take perhaps another 30 years to run. But it is happening. The Demos study of young Londoners shows that they score very high on Tao (or "going with the flow"), Excitement and Independence; these new leading-edge values are, they say, part of a wider shift toward new dominant values in London. But these young Londoners are as attached to core values like Environmental Action, Community and Connectedness as are older people, and "these central, core values are likely to remain central to Londoners over the next 20 years".[15] This means a major new constellation of political attitudes: notions that today seem to belong on the radical fringe will soon be in the middle-aged mainstream, and they suggest that self-sustaining autonomy is an idea whose time is just about to come.

In fact, we know that the Farmers' Market idea is a coming idea, because such markets are thriving in the United States. They are a re-invention of the tradition eroded by prepackaging, long-distance transportation and the hypermarket, and a renewal of the ancient habit of producers coming to town on market days with their fruit and vegetables. They are to be found all over the United States and have different names and habits in different places. They are called Community Farm

---

[14] Thirsk 1997, 263.
[15] Jupp and Lawson 1997, 18.

Markets in Illinois, Food Fairs in Alabama, and Curb Markets in the North-East. In California there used to be only one public market where farmers could sell their produce direct to consumers. By 1990 there were over 50 in that state. The philosopher Marcus Singer, from Madison, Wisconsin, explained its significance to us.

> They emerged from the alternative culture of the sixties and the ambition to grow organically on land where regular farmers had failed. These people come into town in the early morning to set up their stalls with fruit and vegetables that aren't commercially sold. Every Saturday they take over the vast square opposite the state capital building and traffic is diverted. Elsewhere, they use the supermarket car park twice a week and the supermarket actually buys up the produce left at the end of the day. It is still fresher and cheaper.[16]

The Farmers' Market is bound to spread to Britain. The British equivalent has taken the form of box schemes, where a network of families seeking organically grown food agree to take a weekly box of fruit and vegetables in season from the handful of local growers and even allotment-holders. But in September 1997, Britain's first American-style Farmer's Market was held near the old Green Park Station in Bath. It was initiated by Patricia Tutt, the Local Agenda 21 co-ordinator for Bath and North East Somerset, with 40 sellers from a 35-mile (56 km) radius of the city. As we sent this book to the publisher, she told us that by the time of the third Farmer's Market there, over 40 other local authorities had expressed an interest.

One thing that is certain is that the preoccupation with environmental issues, with Agenda 21, and with reducing the demand for private car transport is bound to increase in the early twenty-first century. The proposals which incorporate these themes into plans for the new settlements are those which will be likeliest to win support from that vast franchise of citizens wanting to be Green but never quite sure which way to go.

Howard's century-old prescription remains extraordinarily useful, both for planning policy and for its opponents. The essence of successful political action is surfing: to go with the tide, but to catch the wave in advance. We have that chance now, as Howard had a century ago, but we have to seize the moment. We can build coalitions that will allow us to create twenty-first century versions of Howard's nineteenth-century ideal. It will not be easy, for the forces of superficial prejudice and self-interest are rampant, and they masquerade in the clothing of environmentalism, with which they have nothing to do. But it can be done, and in the interests of all of us, town-dwellers and country-dwellers alike, it *must* be done.

---

[16] Ward 1992.

List of Figures and Tables

Figure 1	Ebenezer Howard	5
Figure 2	Ebenezer Howard Plaque	6
Figure 3	Colonel William Light	13
Figure 4	Adelaide and North Adelaide	14
Figure 5	The Three Magnets, 1898	18
Figure 6	Garden City	20
Figure 7	Social City	24
Figure 8	The Vanishing Point of Landlord's Rent	27
Figure 9	Raymond Unwin	33
Figure 10	Letchworth	37
Figure 11	Letchworth	38
Figure 12	Frederic Osborn	43
Figure 13	Welwyn Garden City	48
Figure 14	Welwyn Garden City	49
Figure 15	Patrick Abercrombie	50
Figure 16	The Greater London Plan 1944	51
Figure 17	The British New Towns	54
Figure 18	Harlow	58
Figure 19	Basildon	60
Figure 20	Cumbernauld	61
Figure 21	Runcorn	62
Figure 22	Milton Keynes	64
Figure 23	Milton Keynes	65
Figure 24	Milton Keynes	66
Figure 25	Plotlands Map	73

212

Figure 26   Laindon   76

Figure 27   Land Settlement – Before   84

Figure 28   Land Settlement – After   84

Figure 29   Copenhagen: The Finger Plan   92

Figure 30   Stockholm: The Markelius Plan   94

Figure 31   Vällingby   95

Figure 32   Paris: The 1965 Strategy   97

Figure 33   Evry   98

Figure 34   The Three Magnets, 1998   104

Figure 35   Household Projections   112

Figure 36   Regional Household Projections   113

Figure 37   Backland Development   127

Figure 38   The Piggeries   132

Figure 39   London RingRail and New Development Clusters   137

Figure 40   Regional Metro and New City Clusters   141

Figure 41   Breheny and Rookwood's Sustainable Development   147

Figure 42   Calthorpe's Transit-Oriented Development   148

Figure 43   Thames Gateway   156

Figure 44   Barking Reach   159

Figure 45   Eastern Quarry   159

Figure 46   The South East Planning Council's 1967 Strategy   161

Figure 47   City of Mercia   165

Figure 48   City of Anglia   167

Figure 49   City of Kent   169

Figure 50   Lightmoor   194

Figure 51   Lightmoor   194

Figure 52   Single-Person Dwelling   197

Figure 53   "Silkingrad"   205

Table 1   Population changes arising from within-Britain migration, 1990–91, by district types   107

Table 2   EU Set-Aside land, 1995   108

REFERENCES

Aalen, F.H.A. (1992) English Origins. In: Ward, S.V. (ed.) *The Garden City: Past, Present and Future*, 28–51. London: Spon.

Abercrombie, P. (1945) *Greater London Plan 1944*. London: HMSO.

Adams, T. (1905) *Garden City and Agriculture: How to Solve the Problem of Rural Depopulation*. London: Simkin Marshall.

Ambrose, P. (1992) The Rural/Urban Fringe as Battleground. In: Short, B. (ed.) *The English Rural Community: Image and Analysis*, 186. Cambridge: Cambridge University Press.

Anon. (1992) New Town Legacies. *Town and Country Planning*, **61**, 298–302.

Anon. (1995) Paul Delouvrier 1914–1995: Le Grand Aménageur de l'Ile-de-France. *Cahiers de l'Institut d'Aménagement et d'Urbanisme de la Région Ile-de-France*, **108**, special supplement.

Anon. (1996) The New Town Experience. *Town and Country Planning*, **65**, 302–304.

Anon. (1997a) Necessity Breeds Ingenuity. *Squall*, **15**, Summer.

Anon. (1997b) Living Within the Social City Region: An Edited Version of the TCPA's Response to the Household Growth: Where Shall We Live? Green Paper. *Town and County Planning*, **66**, 80–82.

Ashworth, W. (1954) *The Genesis of British Town Planning: A Study in Economic and Social History of the Nineteenth and Twentieth Centuries*. London: Routledge & Kegan Paul.

Bailey, J. (1955) *The British Co-operative Movement*. London: Hutchinson's University Library.

Banham, R., Barker, P., Hall, P. and Price, C. (1969) Non-Plan: An Experiment in Freedom. *New Society*, **26**, 435–443.

Banister, D. (1992) Energy Use, Transportation and Settlement Patterns. In: Breheny, M.J. (ed.) *Sustainable Development and Urban Form* (European Research in Regional Science, 2), 160–181. London: Pion.

Banister, D. (1993) Policy Responses in the U.K. In: Banister, D. and Button, K. (eds.) *Transport, the Environment and Sustainable Development*, 53–78. London: Spon.

Banister, D. and Banister, C. (1995) Energy Consumption in Transport in Great Britain: Macro Level Estimates. *Transportation Research, A: Policy and Practice*, **29**, 21–32.

Banister, D. and Button, K. (1993) Environmental Policy and Transport: An Overview. In: Banister, D. and Button, K. (eds.) *Transport, the Environment and Sustainable Development*, 1–15. London: Spon.

Banister, D. and Hall, P. (1995) Summary and Conclusions. In: Banister, D. (ed.) *Transport and Urban Development*, 278–287. London: Spon.

Batchelor, P. (1969) The Origin of the Garden City Concept of Urban Form. *Journal of the Society of Architectural Historians*, **28**, 184–200.

Beevers, R. (1988) *The Garden City Utopia: A Critical Biography of Ebenezer Howard*. London: Macmillan.

Bendixson, T. and Platt, J. (1992) *Milton Keynes: Image and Reality*. Cambridge: Granta Editions.

Benevolo, L. (1967) *The Origins of Modern Town Planning*. Cambridge, MA: MIT Press.

214

Bibby, P. and Shepherd, J. (1997) Projecting Rates of Urbanisation in England, 1991–2016. *Town Planning Review*, **68**, 93–124.

Body, R. (1993) Countryside Planning. In: *Town & Country Planning Summer School, Lancaster*. London: Royal Town Planning Institute.

Bonner, A. (1970) *British Co-operation: The History, Principles and Organisation of the British Co-operative Movement*. Manchester: Co-operative Union.

Bramley, G. (1993) Planning, the Market and Private Housebuilding. *The Planner*, **79/1**, 14–16.

Bramley, G., Bartlett, W. and Lambert, C. (1995) *Planning, the Market and Private Housebuilding*. London: UCL Press.

Breheny, M. (1990) Strategic Planning and Urban Sustainability. In: *Proceedings of TCPA Annual Conference, Planning for Sustainable Development*, 9.1–9.28. London: Town and Country Planning Association.

Breheny, M. (1991) Contradictions of the Compact City. *Town and Country Planning*, **60**, 21.

Breheny, M. (1992) The Contradictions of the Compact City: A Review. In: Breheny, M.J. (ed.) *Sustainable Development and Urban Form* (European Research in Regional Science, 2), 138–159. London: Pion.

Breheny, M. (1995a) *Counter-Urbanisation and Sustainable Urban Forms*. In: Brotchie, J.F., Batty, M., Blakely, E., Hall, P. and Newton, P. (eds) *Cities in Competition*, 402–429. Melbourne: Longman Australia.

Breheny, M. (1995b) The Compact City and Transport Energy Consumption. *Transactions of the Institute of British Geographers*, **20**, 81–101.

Breheny, M. (1995c) Transport Planning, Energy and Development: Improving Our Understanding of the Basic Relationships. In: Banister, D. (ed.) *Transport and Urban Development*, 89–95. London: Spon.

Breheny, M. (1997) Urban Compaction: Feasible and Acceptable? *Cities*, **14**, 209–218.

Breheny, M. and Hall, P. (1996a) Four Million Households – Where Will They Go? *Town and Country Planning*, **65**, 39–41.

Breheny, M. and Hall, P. (1996b) Where Will They Go? A Response to the Response. *Town and Country Planning*, **65**, 290–291.

Breheny, M. and Hall, P. (eds) (1996c) *The People – Where Will They Go? National Report of the TCPA Regional Inquiry into Housing Need and Provision in England*. London: Town and Country Planning Association.

Breheny, M. and Rookwood, R. (1993) Planning the Sustainable City Region. In: Blowers, A. (ed.) *Planning for a Sustainable Environment*, 150–189. London: Earthscan.

Breheny, M., Gent, T. and Lock, D. (1993) *Alternative Development Patterns: New Settlements*. London: HMSO.

Broome, J. and Richardson, B. (1991) *The Self-Build Book: How to Enjoy Designing and Building Your Own House*. Hartland, Devon: Green Books.

Brotchie, J.F., Anderson, M. and McNamara, C. (1995) Changing Metropolitan Commuting Patterns. In: Brotchie, J.F., Batty, M., Blakely, E., Hall, P. and Newton, P. (eds.) *Cities in Competition*. Melbourne: Longman Australia.

Calthorpe, P. (1993) *The Next American Metropolis: Ecology, Community, and the American Dream*. Princeton: Princeton Architectural Press.

Campbell of Eskan, J. (1978) The Future of the Town and Country Planning Association (Address to TCPA AGM 23 May 1978). In: Darley, G. (ed.) (1991) *Tomorrow's New Communities*. York: Joseph Rowntree Foundation.

Caulton, J. (1996) Going to Town on Housing. *Planning Week*, 18 January, 14–15.

Cervero, R. (1985) *Suburban Gridlock*. New Brunswick: Rutgers University, Center for Urban Policy Studies.

Cervero, R. (1989) *America's Suburban Centers: The Land Use–Transportation Link*. Boston: Unwin Hyman.

Cervero, R. (1991) Congestion Relief: The Land Use Alternative. *Journal of Planning Education and Research*, **10**, 119–129.

Champion, A. and Atkins, D. (1996) *The Counterurbanisation Cascade: An Analysis of the Census Special Migration Statistics for Great Britain*. Newcastle upon Tyne: University of Newcastle upon Tyne, Department of Geography.

Cheshire, P.C. (1994) A New Phase of Urban Development in Western Europe? The Evidence for the 1980s. *Urban Studies*, **32**, 1045–1063.

Cheshire, P.C. and Hay, D.G. (1989) *Urban Problems in Western Europe: An Economic Analysis*. London: Unwin Hyman.

Clark, C. (1957) Transport: Maker and Breaker of Cities. *Town Planning Review*, **28**, 237–250.

Collings, T. (ed.) (1987) *Stevenage 1946–1986: Images of the First New Town*. Stevenage: SPA Books.

Commission of the European Communities (1990) *Green Paper on the Urban Environment* (EUR 12902). Brussels: CEC.

Cox, A. (1984) *Adversary Politics and Land: The Conflict over Land and Property Policy in Post-War Britain*. Cambridge: Cambridge University Press.

Craig, D. (1990) *On the Crofters' Trail: In Search of the Clearance Highlanders*. London: Cape.

Creese, W.L. (1966) *The Search for Environment: The Garden City Before and After*. New Haven: Yale University Press.

Cullingworth, J.B. and Nadin, U. (1997) *Town and Country Planning in Britain: Twelfth Edition* (The New Local Government Series, No. 8). London: Routledge.

Culpin, E.G. (1913) *The Garden City Movement Up-to-Date*. London: Garden Cities and Town Planning Association.

Daniels, P.W. and Warnes, A.M. (1980) *Movement in Cities: Spatial Perspectives in Urban Transport and Travel*. London: Methuen.

Darley, G. (1975) *Villages of Vision*. London: Architectural Press.

Darley, G. (ed.) (1991) *Tomorrow's New Communities*. York: Joseph Rowntree Foundation.

Delafons, J. (1990) *Development Impact Fees and Other Devices*. Berkeley: University of California at Berkeley, Institute of Urban and Regional Development (Monograph 40).

Denmark. Egnsplankonteret (1947) *Skitseforslag til Egnsplan for Storkøbenhavn*. Copenhagen: Teknisk Kontor for Udvalget til Planlaegning af Københavnsegnen.

Direction Régionale de l'Equipment d'Ile-de-France (1990) *Les transports de voyageurs en Ile-de-France, 1989*. Paris: DREIF.

Dix, G. (1978) Little Plans and Noble Diagrams. *Town Planning Review*, **49**, 329–352.

Douglas, R. (1976) *Land, People and Politics: A History of the Land Question in the United Kingdom 1978–1952*. London: Allison & Busby.

Ealing Garden Tenants (1912) *The Pioneer Co-Partnership Suburb: A Record of Progress*. London: Co-Partnership Publishers.

Evans, A.W. (1991) Rabbit Hutches on Postage Stamps: Planning, Development and Political Economy. *Urban Studies*, **28**, 853–870.

Evenson, N. (1979) *Paris: A Century of Change, 1878–1978*. New Haven: Yale University Press.

Fairlie, S. (1996) *Low Impact Development: Planning and People in a Sustainable Countryside*. Oxford: Jon Carpenter.

Fehl, G. (1983) The Niddatal Project – The Unfinished Satellite Town on the Outskirts of Frankfurt. *Built Environment*, **9**, 185–197.

Fishman, R. (1977) *Urban Utopias in the Twentieth Century: Ebenezer Howard, Frank Lloyd Wright and Le Corbusier*. New York: Basic Books.

Fulford, C. (1998) *Urban Housing Capacity and the Sustainable City, 1: The Costs of Reclaiming Derelict Sites*. London: Town and Country Planning Association.

Gallion, A.B. and Eisner, S. (1963) *The Urban Pattern*. Princeton: D. van Nostrand.

Garreau, J. (1991) *Edge City: Life on the New Frontier*. New York: Doubleday.

GB Department of the Environment (1969) *People and Planning: Report of the Committee on Public Participation in Planning* (Chairman: A.M. Skeffington). London: HMSO.

216

GB Department of the Environment (1992a) *Planning Policy Guidance: Housing* (PPG 3 (revised)). London: HMSO.

GB Department of the Environment (1992b) *The Relationship between House Prices and Land Supply.* (By Gerald Eve, Chartered Surveyors with the Department of Land Economy, University of Cambridge.) London: HMSO.

GB Department of the Environment (1993a) *East Thames Corridor: A Study of Developmental Capacity and Potential.* By Llewelyn-Davies, Roger Tym and Partners, TecnEcon and Environmental Resources Ltd. London: DOE.

GB Department of the Environment (1993b) *Strategic Planning Guidance for the South East.* London: DOE.

GB Department of the Environment (1995) *Urbanization in England: Projections 1991–2016.* London: HMSO.

GB Department of the Environment and Department of Transport (1993) *Reducing Transport Emissions through Planning* (ECOTEC Research and Consulting Ltd in Association with Transportation Planning Associates.) London: HMSO.

GB Department of the Environment, Transport and the Regions (1998a) *The Future of Regional Planning Guidance: Consultation Paper.* London: DETR.

GB Department of the Environment, Transport and the Regions (1998b) *Modernising Planning: A Policy Statement by the Minister for the Regions, Regeneration and Planning.* London: DETR.

GB Department of the Environment and Welsh Office (1994) *Planning Policy Guidance: Transport (PPG 13).* London: HMSO.

GB Department of Health (1998) *The Quantification of the Effects of Air Pollution on Health in the United Kingdom.* London: HMSO.

GB Deputy Prime Minister and Secretary of State for the Environment, Transport and the Regions (1998) *Planning for the Communities of the Future.* London: HMSO.

GB Government Office for London (1995) *Strategic Guidance for London Planning Authorities: Consultation Draft* (RPG 3). London: GOL.

GB Ministry of Housing and Local Government (1964) *The South East Study 1961–81.* London: HMSO.

GB Secretary of State for the Environment (1996) *Household Growth: Where Shall We Live?* (Cm 3471). London: HMSO.

GB South East Economic Planning Council (1967) *A Strategy for the South East: A First Report by the South East Planning Council.* London: HMSO.

GB South East Joint Planning Team (1970) *Strategic Plan for the South East: A Framework. Report by the South East Joint Planning Team.* London: HMSO.

GB Thames Gateway Task Force (1995) *The Thames Gateway Planning Framework* (RPG 9a). London: Department of the Environment.

Girouard, M. (1985) *Cities and People: A Social and Architectural History.* New Haven: Yale University Press.

Gold, J.R. and Gold, M.M. (1982) Land Settlement Policy in the Scottish Highlands. *Nordia,* **16**, 129–133.

Goodwin, P. and Hass-Klau, C. (1996) *The Real Effects of Environmentally Friendly Transport Policies.* Summary Report. (Mimeo.)

Gordon, P. and Richardson, H.W. (1989) Gasoline Consumption and Cities – A Reply. *Journal of the American Planning Association,* **55**, 342–346.

Gordon, P., Richardson, H.W. and Jun, M. (1991) The Commuting Paradox – Evidence from the Top Twenty. *Journal of the American Planning Association,* **57**, 416–420.

Green, F.E. (1912) *The Awakening of England.* London: Thomas Nelson.

Haggard, H.R. (1905) *The Poor and the Land.* London: Longmans, Green.

Hall, P. (1967) Planning for Urban Growth: Metropolitan Area Plans and Their Implications for South-East England. *Regional Studies,* **1**, 101–134.

Hall, P. (1988) *Cities of Tomorrow: An Intellectual History of Urban Planning and Design in the Twentieth Century.* Oxford: Basil Blackwell.

Hall, P. (1992) *Urban and Regional Planning,* Third Edition. London: Routledge.

Hall, P. (1995) A European Perspective on the Spatial Links between Land Use, Development and Transport. In: Banister, D. (ed.) *Transport and Urban Development*, 65–88. London: Spon.

Hall, P. (1996a) 1946–1996 – From New Town to Sustainable Social City. *Town and Country Planning*, **65**, 295–297.

Hall, P. (1996b) Le New Towns in Gran Bretagna: Passato, Presente e Futuro. *Urbanistica*, **107**, 141–145.

Hall, P. and Banister, D. (1995) Summary and Conclusions. In: Banister, D. (ed.) *Transport and Urban Development*, 278–287. London: Spon.

Hall, P. and Hay, D. (1980) *Growth Centres in the European Urban System*. London: Heinemann.

Hall, P., Sands, B. and Streeter, W. (1993) *Managing the Suburban Commute: A Cross-National Comparison of Three Metropolitan Areas*. Berkeley: University of California at Berkeley, Institute of Urban and Regional Development, Working Paper 596.

Handy, S.L. and Mokhtarian, P.L. (1995) Planning for Telecommuting: Measurement and Policy Issues. *Journal of the American Planning Association*, **61**, 99–111.

Hardy, D. (1979) *Alternative Communities in Nineteenth Century England*. London: Longman.

Hardy, D. (1989) War, Planning and Social Change: The Example of the Garden City Campaign, 1914–1918. *Planning Perspectives*, **4**, 187–206.

Hardy, D. (1991a) *From Garden Cities to New Towns: Campaigning for Town and Country Planning, 1899–1946*. London: Spon.

Hardy, D. (1991b) *From New Towns to Green Politics: Campaigning for Town and Country Planning, 1946–1990*. London: Spon.

Hardy, D. and Ward, C. (1984) *Arcadia for All: The Legacy of a Makeshift Landscape*. London: Mansell.

Hartmann, K. (1976) *Deutsche Gartenstadtbewegung: Kulturpolitik und Gesellschaftsreform*. Munich: Heinz Moos Verlag.

Headicar, P. and Curtis, C. (1996) *The Location of New Residential Development: Its Influence on Car-Based Travel* (Oxford Planning Monograph, 1/2). Oxford: Oxford Brookes University, School of Planning.

Hebbert, M. (1992) The British Garden City: Metamorphosis. In: Ward, S.V. (ed.) *The Garden City: Past, Present and Future*, 165–186. London: Spon.

Hedges, B. and Clemens, S. (1994) *Housing Attitudes Survey*. London: HMSO.

Housing Research Foundation (1998) *Homealone: The Housing Preferences of One Person Households*. Amersham: Housing Research Foundation.

Howard, E. (1898) *To-morrow! A Peaceful Path to Real Reform*. London: Swan Sonnenschein.

Howard, E. (1902) *Garden Cities of To-Morrow*. London: Swan Sonnenschein.

Howard, E. (1946) *Garden Cities of Tomorrow*. London: Faber & Faber.

Howard, E. (1904) opening the discussion of a paper by Patrick Geddes at the London School of Economics, reprinted in Meller, H. (1979) *The Ideal City*. Leicester: Leicester University Press.

Hughes, M. (ed.) (1971) *The Letters of Lewis Mumford and Frederic J. Osborn: A Transatlantic Dialogue 1938–70*. New York: Praeger.

Institut d'Aménagement et d'Urbanisme de la Région d'Ile-de-France, and Conseil Régional Ile-de-France (1990) *ORBITALE: Un Réseau de Transports en Commun de Rocade en Zone Centrale*. Paris: IAURIF.

Jackson, K.T. (1985) *Crabgrass Frontier: The Suburbanization of the United States*. New York: Oxford University Press.

Jupp, B. and Lawson, G. (1997) *Values Added: How Emerging Values Could Influence the Development of London*. London: Demos.

Kampffmeyer, H. (1908) Die Gartenstadtbewegung. *Jahrbücher für Nationalökonomie und Statistik, III. Serie*, **36**, 577–609.

Kelbaugh, D. et al. (eds) (1989) *The Pedestrian Pocket Book: A New Suburban Design*

Strategy. New York: Princeton Architectural Press in association with the University of Washington.

Kenworthy, J., Laube, F., Newman, P. and Barter, P. (1997) *Indicators of Transport Sustainability in 37 Global Cities: A Report for the World Bank.* Perth: Murdoch University, Institute for Science and Technology Policy, Sustainable Transport Research Group.

King, A.D. (1984) *The Bungalow: The Production of a Global Culture.* London: Routledge & Kegan Paul.

Kohr, L. (1957) *The Breakdown of Nations.* London: Routledge & Kegan Paul.

Lane, B.M. (1968) *Architecture and Politics in Germany, 1918–1945.* Cambridge, MA: Harvard University Press.

Leneman, L. (1989) *Fit for Heroes? Land Settlement in Scotland after World War One.* Aberdeen: Aberdeen University Press.

Llewelyn-Davies Planning (1997) *Sustainable Residential Quality: New Approaches to Urban Living.* London: Llewelyn-Davies.

Lock, D. (1995) Room for More Within the City Limits? *Town and Country Planning,* **67,** 173–176.

London Planning Advisory Committee (1993) *Draft 1993 Advice on Strategic Planning Guidance for London.* London: LPAC.

London Transport (1995) *Planning London's Transport.* London: LT.

London Transport (1996) *Planning London's Transport: To Win as a World City.* London: LT.

Macfadyen, D. (1933) *Sir Ebenezer Howard and the Town Planning Movement.* Manchester: Manchester University Press.

MacKenzie, N. and MacKenzie, J. (1977) *The Fabians.* New York: Simon & Schuster.

Mackett, R. (1993) *Why Are Continental Cities More Civilised than British Ones?* Paper presented at 25th Universities Transport Studies Group Annual Conference, Southampton, January.

Marsh, J. (1982) *Back to the Land: The Pastoral Impulse in England from 1880 to 1914.* London: Quartet Books.

Marshall, A. (1884) The Housing of the London Poor. I. Where to House Them. *The Contemporary Review,* **45,** 224–231.

McCready, K.J. (1974) *The Land Settlement Association: Its History and Present Form.* London: Plunkett Foundation for Co-operative Studies.

Meyerson, M. (1961) Utopian Traditions and the Planning of Cities. *Daedalus,* 90/1, 180–193.

Miller, M (1983) Letchworth Garden City Eighty Years On. *Built Environment,* 9, 167–184.

Miller, M. (1989) *Letchworth: The First Garden City.* Chichester: Phillimore.

Ministry of Agriculture, Fisheries and Food (1996) *The Digest of Agricultural Census Statistics, United Kingdom, 1995.* London: HMSO.

Mokhtarian, P.L. (1991) Telecommuting and Travel: State of the Practice, State of the Art. *Transportation,* **18,** 319–342.

Moss-Eccardt, J. (1973) *Ebenezer Howard.* Tring: Shire Publications.

Mullin, J.R. and Payne, K. (1997) Thoughts on Edward Bellamy as City Planner: The Ordered Art of Geometry. *Planning History Studies,* **11,** 17–29.

Mumford, L. (1946) The Garden City Idea and Modern Planning. In: Howard, E., *Garden Cities of Tomorrow,* 29–40. London: Faber & Faber.

Netherlands, Ministry of Housing, Physical Planning and the Environment (1991) *Fourth Report (EXTRA) on Physical Planning in the Netherlands: Comprehensive Summary: On the Road to 2015.* The Hague: Ministry of Housing, Physical Planning and the Environment, Department for Information and International Relations.

Newman, P.W.G. and Kenworthy, J.R. (1989a) *Cities and Automobile Dependence: A Sourcebook.* Aldershot and Brookfield, VT: Gower.

Newman, P.W.G. and Kenworthy, J.R. (1989b) Gasoline Consumption and Cities: A

Comparison of U.S. Cities with a Global Survey. *Journal of the American Planning Association*, **55**, 24–37.

Newman, P.W.G. and Kenworthy, J.R. (1992) Is There a Role for Physical Planners? *Journal of the American Planning Association*, **58**, 353–362.

Newman, P.W.G. and Kenworthy, J.R. (1997) *Sustainability and Cities*. Washington, DC: Island Press.

Newman, P.W.G., Kenworthy, J.R. and Laube, F. (1998) The Global City and Sustainability: Perspectives from Australian Cities and a Survey of 37 Global Cities. In: Brotchie. J.R., Dickey, J., Hall, P. and Newton, P. (eds) *East–West Perspectives on 21st Century Urban Development: Sustainable Asian and Western Cities in the New Millennium*. Melbourne: Longman Australia (forthcoming).

Osborn, F.J. (1946) Preface. In: Howard, E., *Garden Cities of Tomorrow*, 9–28. London: Faber & Faber.

Osborn, F.J. (1950) Sir Ebenezer Howard: The Evolution of His Ideas. *Town Planning Review*, **21**, 221–235.

Osborn, F.J. (1970) *Genesis of Welwyn Garden City: Some Jubilee Memories*. London: Town and Country Planning Association.

Osborn, F.J. and Whittick, A. (1963) *The New Towns: The Answer to Megalopolis*. London: Leonard Hill.

Owens, S.E. (1984) Spatial Structure and Energy Demand. In: Cope, D.R., Hills, P.R. and James, P. (eds) *Energy Policy and Land Use Planning*, 215–240. Oxford: Pergamon.

Owens, S.E. (1986) *Energy, Planning and Urban Form*. London: Pion.

Owens, S.E. (1990) Land-Use Planning for Energy Efficiency. In: Cullingworth, J.B. (ed.) *Energy, Land and Public Policy*, 53–98. Newark, DE: Transactions Publishers, Center for Energy and Urban Policy Research.

Owens, S.E. (1992a) Energy, Environmental Sustainability and Land-Use Planning. In: Breheny, M.J. (ed.) *Sustainable Development and Urban Form* (European Research in Regional Science, 2), 79–105. London: Pion.

Owens, S.E. (1992b) Land-Use Planning for Energy Efficiency. *Applied Energy*, **43**, 81–114.

Owens, S.E. and Cope, D. (1992) *Land Use Planning Policy and Climate Change*. London: HMSO.

Pitt, J. (1995) Building on Resources. *Planning Week*, **3/15** (13 April), 14–16.

Potter, S. (1976) *Transport and New Towns: The Transport Assumptions Underlying the Design of Britain's New Towns 1946–1976*. Milton Keynes: Open University New Towns Study Unit.

Ratcliffe, J. and Stubbs, M. (1996) *Urban Planning and Real Estate Development*. London: UCL Press.

Rave, R. and Knöfel, H-J. (1968) *Bauen seit 1900 in Berlin*. Berlin: Kiepert.

Read, J. (1978) The Garden City and the Growth of Paris. *Architectural Review*, **113**, 345–352.

Rickaby, P.A. (1987) Six Settlement Patterns Compared. *Environment and Planning, B*, **14**, 193–223.

Rickaby, P.A. (1991) Energy and Urban Development in an Archetypal English Town. *Environment and Planning, B*, **18**, 153–176.

Rickaby, P.A., Steadman, J.B. and Barrett, M. (1992) Patterns of Land Use in English Towns: Implications for Energy Use and Carbon Monoxide Emissions. In: Breheny, M.J. (ed.) *Sustainable Development and Urban Form* (European Research in Regional Science, 2), 182–196. London: Pion.

Rural Development Commission (1998) *Young People and Housing*. York: York Publishing Services.

Schipper, L. and Meyers, S. (1992) *Energy Efficiency and Human Activity: Past Trends, Future Prospects* (Cambridge Studies in Energy and the Environment). Cambridge: Cambridge University Press.

SERPLAN (1998) *A Sustainable Development Strategy for the South East: Public Consultation*. London: SERPLAN (SPR400).

Sharp, T. (1932) *Town and Countryside: Some Aspects of Urban and Rural Development*. London: Oxford University Press.

Simpson, M. (1985) *Thomas Adams and the Modern Planning Movement: Britain, Canada and the United States, 1900–1940*. London: Mansell.

Smith, N.R. (1946) *Land for the Small Man*. Morningside Heights, NY: King's Crown Press.

Soria y Pug, A. (1968) *Arturo Soria y la Ciudad Lineal*. Madrid: Revista de Occidente.

Stern, R.A.M. (1986) *Pride of Place: Building the American Dream*. Boston: Houghton Mifflin.

Stone, P.A. (1973) *The Structure, Size and Costs of Urban Settlements* (National Institute of Economic and Social Research, Economic and Social Studies, XXVIII). Cambridge: Cambridge University Press.

Sudjic, D. (1992) *The 100 Mile City*. London: Andre Deutsch.

Sweden, National Rail Administration, City of Stockholm, Stockholm County Council and National Road Administration (1993) *The Dennis Agreement: Traffic System of the Future*. Stockholm: City, Office of Regional Planning and Urban Transportation.

Swenarton, M. (1985) Sellier and Unwin. *Planning History Bulletin*, **7/2**, 50–57.

Tegnér, G. (1994) *The "Dennis Traffic Agreement" – a Coherent Transport Strategy for a Better Environment in the Stockholm Metropolitan Region*. Paper presented at the STOA International Workshop, Brussels, April 1994.

Thacker, J. (1993) *Whiteway Colony: The Social History of a Tolstoyan Community*. Whiteway, Gloucestershire: The Author.

Thirsk, J. (1997) *Alternative Agriculture: A History from the Black Death to the Present Day*. Oxford: Oxford University Press.

Thomas, R. (1969) *London's New Towns: A Study of Self-Contained and Balanced Communities*. London: Political and Economic Planning (PEP).

Thomas, R. (1996) The Economics of the New Towns Revisited. *Town and Country Planning*, **65**, 305–308.

Thomas, W. (1983) *Letchworth: The First Garden City: Celebrating Eighty Years of Progress towards a Better Environment 1903–1983* (The Ebenezer Howard Memorial Lecture). Letchworth: ?Privately Printed.

Thompson, F.M.L. (1965) Land and Politics in England in the Nineteenth Century. *Transactions of the Royal Historical Society*, **15**, 23–44.

Todd, N. (1986) *Roses and Revolutionaries: The Story of the Clousden Hill Free Communist and Co-operative Colony*. London: People's Publications.

Town and Country Planning Association (1997) *Building the New Britain: Finding the Land for 4.4 Million New Households*. London: TCPA.

Trevelyan, G.M. (1937) Amenities and the State. In: Williams-Ellis, C. (ed.) *Britain and the Beast*, 183–186. London: J.M. Dent.

Uhlig, G. (1977) Stadtplanung in den Weimarer Republik: sozialistische Reformaspekte. In: Neue Gesellschaft für Bildende Kunst (ed.) *Wem gehört du Welt? Kunst und Gesellschaft in der Weimarer Republik*. Berlin: Neue Gesellschaft für Bildende Kunst.

UK Round Table on Sustainable Development (1997) *Housing and Urban Capacity*. London: UK Round Table.

van den Berg, L., Drewett, R., Klaassen, L.H., Rossi, A. and Vijverberg, C.H.T. (1982) *Urban Europe: A Study of Growth and Decline* (Urban Europe, Volume 1). Oxford: Pergamon.

Ward, C. (1983) Growing Pains. *New Society*, 20 January.

Ward, C. (1990) *Talking Houses*. London: Freedom Press.

Ward, C. (1993) *New Town, Home Town: The Lessons of Experience*. London: Gulbenkian Foundation.

Ward, C. (1994) Lost in the Global Hypermarket. *New Statesman*, 25 November.

Ward, S.V. (ed.) (1992) *The Garden City: Past, Present and Future*. London: Spon.

Webb, B. (1938) *My Apprenticeship – 2*. Harmondsworth: Penguin Books.

Wells, H.G. (1901) *Anticipations of the Reaction of Mechanical and Scientific Progress upon Human Life and Thought*. London: Chapman & Hall.

Whittick, A. (1987) *F.J.O. – Practical Idealist: A Biography of Sir Frederic J. Osborn*. London: Town and Country Planning Association.

Williams-Ellis, C. (1928) *England and the Octopus*. London: Geoffrey Bles.

Williams-Ellis, C. (ed.) (1937) *Britain and the Beast*. London: J.M. Dent.

Wise, M.J. (Chairman) (1967) *Report of the Departmental Committee of Enquiry into Small Holdings, Part II: Land Settlement*. London: HMSO.

Wood, A. (1988) *Greentown: A Case Study of a Proposed Alternative Community*. Milton Keynes: Open University Energy and Environment Research Unit.

Yago, G. (1984) *The Decline of Transit: Urban Transportation in German and U.S. Cities, (1900–1970)*. Cambridge: Cambridge University Press.

INDEX

Abercrombie, Sir Patrick 47, 49–51, 75–6, 203
Academic wisdom 144–6
Adams, Thomas 30, 82
Addison, Christopher 41
Adelaide 11, 14
Agencies 121, 171–90
Agriculture
  alternative 207–9
  intensification 200
Alexandra Palace 21
Allotment garden 206–7
Ambrose, Peter 199
American Edge Cities 124
Andover 56, 67
Areas of Outstanding Natural Beauty 116, 154
Areas of Tranquility 154
Attlee, Clement 49

Back to the Land movement 12
Backland development 126, 127
BANANAism 111, 121
Banister, David 145, 146
Barking Reach 128, 158, 159
Barnett, Dame Henrietta 41
Basildon 55, 61, 76, 77
Basingstoke 56, 67
Bellamy, Edward 10–11
Benoit-Levy, Georges 88
Bergen 133
Berlin 89–90
Bibby, P. 116
Bland, Hubert 6
Blatchford, Robert 79, 81
Board of Management 28
Body, Sir Richard 200
Booth, William 79
Borer, Pat 197
Bournville 12
Bramley, G. 110
Bramshott 76
Bray, Reginald 75

Breakdown of Nations, The 199
Breheny, Michael 59, 129, 144, 145, 147, 149
Brentham Garden Suburb 206, 207
Britain and the Beast 75
Britz 90
Brotchie, J.F. 145
Brotherhood Church 6
Brownfield land 117
Brownfield town cramming 132
Buckingham, James Silk 12
Building regulations 206
Burnham, Daniel 154
Burns, John 81

Caborn, Dick 115
Cadbury, Edward 32
Cadbury, George 12
Cadbury, L.J. 52
California 146, 148, 187
Calthorpe, Peter 146, 148, 149
Campbell, Lord, of Eskan 192
Capital Gains Tax (CGT) 185, 186, 189–90
Car usage 119, 120, 201–2
Cardiff Bay Development Corporation 182
Central Capital Region 155
Central Lancashire 53
Centralism 179
Chamberlain, Joseph 8
Chamberlain, Neville 47
Champion, Henry 6
Champion, Tony 106
Channel Tunnel Rail Link 140, 141, 153, 156, 168
Chatham Maritime 157
Chelsea–Hackney link 142
Cheshire, Paul 125
Chief Land Registry Tribunal 79
Chubb, Percival 6
Cités Nouvelles 96
City of Anglia 164–7

224

City of Kent 166–8
City of Mercia 162–5
City vs. countryside 17–19
Clark, Colin 155
Clemens, S. 105, 119
Clousden Hill Free Communist and Co-
 operative Colony 80, 81
Clustered development 153
Coalitions, self-sustaining 206–7
Commission for the New Towns (CNT)
 180–1, 183–4
 investing net receipts 186
Community infrastructure investment
 185–6
Community Land Act 1975 174
Compensation and Betterment problem
 204
Conquest of Bread, The 80
Consortium Developments 63
Contemporary practice 146–9
Co-operation 28
Co-operative Land Society 14
Co-operative movement 31, 80
Copenhagen 91, 92
Corby 53, 55
Council for the Preservation of Rural
 England (CPRE) 47, 203
Council for the Protection of Rural England
 199
Counterurbanisation cascade 106
Country Landowners' Association 199
Countryside
 repopulation 106–7
 vs. city 17–19
Countryside Alliance 200
Crofting Act 1886 83
CrossRail 136
Croydon Tramlink 139
Crystal Palace 21, 22
Cumbernauld 53, 56, 57, 61
Curtis, Carey 146
Cwmbran 53

Daniels, Peter 134
Davidson, Thomas 6, 9
Dawley 57
Decentralisation 154
Delafons, John 188
Delouvrier, Paul 96
Density 22
 see also Population density
Density Pyramids 154
Department of the Environment (DOE) 125
Department of the Environment, Transport
 and the Regions (DETR) 126

de Soissons, Louis 49
Developer Contributions 188
Development Agreements 188
Development Corporation 192, 193
Development Impact Fees 187–8
Development Land Tax (DLT) 174,
 183
Development value 25–8
Distance from metropolitan cities 152
Do-it-yourself New Towns 191–8, 203
Dungeness 77

East Kilbride 53
Eastern Quarry 159
Ebbsfleet 156, 159
ECOTEC 125
Electric tramways 71
England and the Octopus 75
English Land Restoration League 7
English Partnerships 182–4
Enterprise Zones 63
Erskine, Ralph 131
European Commission 107
European Union 116
Evans, Alan 109
Eve, Gerald and Partners 109
Examinations in Public (EIPs) 178

Fabian Society 6, 9, 30
Fairlie, Simon 195–6
Family Income Supplement 85
Farmers' Market 208–9
Farming industry 200–1
Farrar, Dean 17
Fellowship of the New Life 6, 9
Fels, Joseph 81–2
Fields, Factories and Workshops 11,
 79
Finance 121
Financial model 25–8
Financial resources 118
Finger Plan 91, 92
First Garden City Company 32, 34, 36
Fiscal restraints 119
Fishman, Robert 32
*Fourth Report (EXTRA) on Physical
 Planning in the Netherlands*
 146–8
Foxley Wood 63
France 88, 96–8
Frankfurt am Main 88
Frensham 76
Funding 171
Future of Regional Planning Guidance
 177

Garden Cities   12, 17–39, 41–2, 55, 143,
    197
    administration   26
    and Town Planning Association   39
    basic notion   20
    population   20, 23
    residential area   22
    town city   21
Garden Cities and Agriculture   82
Garden Cities of To-Morrow   3
Garden City Association   6, 8, 29–39, 202–3
    objectives   29
Garden City Pioneer Company   31, 32
Garden suburbs   41–2
Gartenstädte   88–90
George, Henry   7
Germany   88–90
Glasgow Eastern Area Renewal (GEAR)   63,
    182
Glenrothes   53, 55
Gordon, P.   144
Gorst, Sir John   17, 79
Government Office for London (GOL)   126
'Great outdoors'   72
Greater London Plan   49, 51, 75, 203
Green Belts   116
Green, F.E.   82
Green Paper 1996   114–16
Greenfield development   117, 152
Greentown Group   192, 193, 195
Greenwich, Millennium Village   131
Greenwich Peninsula   128
Grossiedlung   89
Gummer, John   178, 205
Gurneys   4–5

Haggard, Rider   82
Hall, David   193
Hampstead Garden Suburb   35, 41–2, 65
Hardie, Keir   9
Häring, Hugo   90
Harlow   59, 68
Harmsworth, Alfred   32
Harmsworth, Cecil   4
Hatfield   55
Haverhill   67
Havering Park   76
Havering Riverside   128, 158
Headicar, Peter   146
Hebbert, Michael   52, 65, 67, 69
Hedges, B.   105, 119
Hellerau   88
Hemel Hempstead   163
Hertfordshire   130
Heseltine, Michael   181, 182

High-rise blocks   131
High-speed train link   149, 153, 155
Hitchin   55
Hoggart, Richard   71
Holidays With Pay Act 1938   72
Hollesley Bay   81
Holmans, Alan   112
Hook   57
Hooper, Alan   112
House prices   109
Housing   17, 22
    projections   111–14, 131
    regional needs   175–9
    self-build   197
Housing and Planning Act 1909   39
Housing and Planning Act 1986   196
Housing market   110
Housing Research Foundation   119
Housing White Paper 1995   114
Howard, Ebenezer   3–15, 32, 42, 72, 79,
    191
    birthplace plaque   12
Howardsgate Shopping Centre   47
Hufeneisensiedlung   90
Hyndman, H.M.   6, 9, 28

Idris, T.H.W.   32
Impact Fee agreements   186
Improvement Rate   35
In Darkest England, and the Way Out   79
Individualism   28
Industrial villages   12
Inner Urban Areas Act 1978   63
Inter-Municipal Railway   55
International Metro   140
Irish Home Rule   7
Irvine   53, 57
Isle of Sheppey   128

Jenkins, Hugh   128
Jobs   23
Joseph Rowntree Foundation   118

Kampffmeyer, Hans   88
Kapper, Frank   80
Kenworthy, John C.   80
Kenworthy, J.R.   134, 144, 145
Key, William   80
King, Anthony   77
Kingdom of God is Within You, The   79
Kohr, Leopold   199
Kropotkin, Peter   11, 28, 79, 80, 208

La Conquête du Pain   80
La Cuidad Lineal   87

226

Labour colonies 81
Laindon 76, 81
Land Authority 183–4
Land Authority for Wales (LAW) 174, 175, 183, 186
Land colonisation 10, 82
Land Commission Act 1967 173
Land Compensation Act 1961 173
Land development 109
Land development agencies 179
Land Nationalisation 15, 74
Land Nationalisation Society 8, 14, 31
Land prices 72
Land question 7–9
Land reform 8
Land settlement 79–86
Land Settlement Association (LSA) 83
Land Settlement (Facilities) Act 1919 83
Land speculation 36
Land tax 7
Land use planning 142–9
Land values 39
  rural 107–11
Landmann, Ludwig 88
Langdon Hill 75
Lansbury, George 81
Lawson, Nigel 174
Le Cité Jardin 88
Lee Valley Regional Park 129
Legislation and counter-legislation 179
Leighton Buzzard 163
Letchworth 3, 22, 31, 32, 34–6, 55, 152
Lever, William Hesketh 12, 32, 35
Light rail systems 133
Light, William 13
Lightmoor 193–5
Linslade 163
Livingston 53
Llewelyn-Davies Planning (LDP) 115, 126, 128, 130
Local Government, Planning and Land Act 1980 181
Lock, David 130, 131
London 6–7, 23, 26, 49, 51, 75, 126–31, 136–42, 203, 204
London Docklands 129
London Planning Advisory Committee (LPAC) 126, 129, 130
London Regional Metro 153, 162
London Transport 141, 142
Looking Backward 10
LUTECE 135
Lymm 56
Lynch, Kevin 160

Madrid 87
Magnetism 18
Mark One New Towns 56, 59, 105, 134, 143, 153, 163
Mark Two New Towns 57, 59, 105, 152, 153, 163
Mark Three New Towns 57, 59, 175
Markelius–Sidenbladh 1952 plan for Stockholm 117
Markelius, Sven 93, 145
Marne-la-Vallée 98
Marshall, Alfred 10
Marshall, Howard 75
May, Ernst 88–90
Mayland 81
Mellor, David 128
Merrie England 79, 81
Migration 17, 26, 106, 107
Mill, J.S. 7, 9
Millennium Village, Greenwich 131
Milton Keynes 55, 57, 59, 65, 66, 68, 163, 185, 193, 195, 197, 205, 206
Milton Keynes Development Corporation 192
Modernising Planning 177
Morris, William 6, 12, 79
Mumford, Lewis 22–3, 36, 191

National Housing Reform Council 8
National Parks 116
Nationalisation of Labour Society 11
Neighbourhood units 21
Netherlands 146, 146–9
Neville, Ralph 30, 32, 34, 39, 67
New City Clusters 141
New Development Clusters 137
New Town Development Corporations 171, 175, 179–81
New Towns 153
  do-it-yourself 191–8, 203
  lessons learned from 67–9
  map 54
  Mark One 56, 59, 105, 134, 143, 153, 183
  Mark Two 57, 59, 105, 152, 153, 163
  Mark Three 57, 59, 175
  mechanism 190
  postwar 52–67
New Towns Act 1946 47, 52, 173, 204
New Towns (Amendment) Act 1976 180
New York Regional Plan (1929–31) 56
Newbury bypass 202
Newman, P.W.G. 134, 144, 145
News From Nowhere 79
Newton Aycliffe 53
Newtown 53, 55

NIMBYism 107, 110–11, 115, 121, 125–7, 152, 176, 178, 179, 189, 199–209
Non-Plan experiment 191
Norris, Steven 201–2
North Circular 138, 139
North Tyneside Youth Self Build Enterprise 197
Northampton 57, 59, 163, 180
Northfield, Lord 193
Nothing Gained by Overcrowding! 130

Olmsted, Frederick Law 4
Onkel-Toms-Hütte 90
ORBITALE, Paris 134–9
Osborn, F.J. 3, 29, 42–5, 47, 52, 53, 55, 59, 191, 203
Oslo 133
Owens, Susan 145

Paris 96–8
ORBITALE 134–9
Parker, Barry 22, 32, 36, 42, 56, 89
Parking 130, 146
Parks 129
Peacehaven 71
Pearsall, Howard 32
Pease, Edward 6, 30
Perry, Clarence 21, 56
Peterborough 57, 59, 152, 180
Peterlee 53
Pitsea 75, 76
Planned cities 11–14
Planning, public participation in 201
Planning Agreements 186, 188
Planning Alliance 205
Planning and Compensation Act 1991 186
Planning effects 110
Planning for the Communities of the Future 177, 178
Planning Front 205
Planning gain 186–7, 203
Planning guidance 188
Planning legislation 76
Planning obligations 186–7, 187
Planning policy 109
Planning Policy Guidance 176
Planning system 173, 175, 187, 203
Playing fields 129
Plessy-Robinson 206
Plotlands 71–7, 206
    attitudes towards 77
    characteristics 74
    map 73
Pointe Gourde system 173, 183
Policy formulation 57

Pollution 130, 144
Poole, Willie 200
Population changes 107
Population density 22, 105, 109, 130, 151
Port Sunlight 12
Portmerrion 75
Potter, Stephen 202
Potters Bar 130
Price of rural land 72
Principles of Political Economy 9
Priority Areas for Regeneration and Development (PARDs) 176, 185, 190
Private enterprise 171
Progress and Poverty 7
Property ownership 74
Protesters 202
Public participation in planning 201
Public transport 119, 131–42, 146
Purdom, C.B. 42

Railways 55, 72, 96, 128, 133, 136–41, 149, 152–3, 155, 156, 168
Rainham Marshes 157
Ramuz, Frederick Francis 74
Rapid transit system 23
Rates 26
Raymond Unwin paradox 130
Raynsford, Nick 107, 177
Redditch 53, 57
Regional Development Agencies 176, 185
Regional express rail (RER) system 96
Regional Guidance 176–7, 183
Regional Land Agencies 185, 189
Regional Land Authorities 184, 186
Regional Metro 139–42
Regional Offices 176, 177
Regional Planning Guidance 111, 125
Regional Standing Conferences 176
Regional Structural Plan 177
Reith Committee 110
Reith, Sir John 52
Religious Society of Friends 83
Remote rural areas 154
Rents 26
Richardson, H.W. 144
Ride-sharing 119
RingRail 136–9
Ritzema, Thomas Purvis 32
Road investments 133
Road pricing 133, 134
Road protesters 201–2
Rochdale Society 31
Römerstadt 206, 207
Rookwood, Ralph 145, 147, 149
Rosehaugh Stanhope 129

228

Round Table on Sustainable Development
115
Royal Commission on Local Taxation 8
Royal Docks 128, 157
Runcorn 53, 57, 63
Rural Development Commission 201
Rural land
price of 72
value 107–11
Ruskin, John 12

Sacramento 146
Salisbury, Lord 17
San Francisco Bay Area of California 187
Satellites 143
Scandinavia 91–5, 143
Scotland 83
Segal, Walter 197
Self-build housing 197
Self-sustaining coalitions 206–7
Sellier, Henri 88, 96
Selsey Peninsula 85
Sequential test 117
SERPLAN 130, 178
Set-aside land 107, 116
Sharp, Thomas 71
Shaw, George Bernard 6, 30, 34, 39
Shepherd, J. 116
Shopping 21, 39, 160
Sidenbladh, Göran 93, 145
Siedlung 89
Siemensstadt 89–90
Silicon Valley 146
Silkin, Lewis 49, 52, 56
SILKINGRAD 56, 205
Simplified Planning Zones (SPZs) 196
Single-person dwelling 197
Sinking-fund 26
Skeffington Committee 201
Skelmersdale 53, 57
Small Holdings and Allotments Act 1908 82
Smallholdings 82, 207
Smith, Thomas 81
Social City 12, 23, 24, 55, 143
Social City Region concept 121
Social Democratic Foundation 9
Socialism 28
Soria y Mata, Arturo 87
South East Economic Planning Council
160, 161
South East Study 174
South Woodham Ferrers 76
Spain 87
Special Development Order 181
Spence, Thomas 9

Spencer, Herbert 7, 9
Steere, F.W. 31
Stevenage 52, 55, 205
Stewart, Sir Malcolm 52
Stockholm 91, 93, 94, 143, 145
Markelius–Sidenbladh 1952 plan 117
Stonehouse 57
Strategic design implementation 168–70
Strategic Plan for the South East 152–4,
161
Strategic policy elements 151–70
Stratford 156, 157
Structure/District Plan system 177
Sub-Regional New Town Development
Corporations 184
Subsidies 118
Sustainability 23, 121, 123–49
challenge to 126–31
Sustainable communities 121–2
Sustainable development 147
Sustainable Development Corridors 162–70
Sustainable growth 111–15
Sustainable Social City Region 120
Sustainable urban development 117, 123
Swindon 56

Taut, Bruno 90
Taylor, Frank 67
Taylor, W.G. 42
Taylor Woodrow 52
Telford 53, 57, 197, 206
Telford Development Corporation 193
Thames Gateway 128, 138, 149, 154–60
Thames Gateway Metro 149
Thames Gateway Planning Framework
155
Thameslink 2000 140, 153, 163, 168
Thamesside Kent 158
Thirsk, J. 207, 208
Thomas, Ray 59, 68
Thomas, Wyndham 180
Thomasson, Franklin 32
Three Magnets diagram 17–19, 28, 79,
103–22, 143, 201
Tolstoy, Leo 79
To-Morrow! A Peaceful Path to Real
Reform 3
Town and Country Planning Act 1932 52
Town and Country Planning Act 1947 108,
172, 204
Town and Country Planning Act 1990
110
Town and Country Planning Association
(TCPA) 47, 53, 59, 110, 114, 115, 118,
120, 177, 178, 184, 192, 203

Town Development Act 1952 56, 175
Town expansions 153
Trabanten 89
Trabantenstädte 88
Traffic calming 120
Traffic congestion 130
Traffic reduction strategies 119
Tramways 71, 87, 133
Transit-Oriented Development (TOD) 146, 148
Transport 2000 136
Transport policies 72, 131–4
Trevelyan, G.M. 75
Trondheim 133
Tunnelbana system 93, 124, 143, 145

Underground railways 128
Unilateral Undertaking 186
Unitary Development Plans 177
Unwin, Raymond 22, 32, 33, 35, 36, 41, 42, 67, 89
Urban compaction 151
Urban Development Corporations (UDCs) 63, 171, 175, 181–2
Urban green spaces 151
Urban greenfield 128, 129
Urban management 124
Urban nodes 151
Urban quality 151
Urban road pricing 133, 134
Uthwatt Committee report (1942) 187

Vällingby 95
Vanishing Point of Landlord's Rent 26

Värby Gärd, Stockholm 131
Variation according to geography 154
Villes Nouvelles 96

Wagner, Martin 89, 90
Wakefield, Edward Gibbon 9, 11
Waldsiedlung 90
Wales 174, 175, 183, 186
Wallace, Alfred Russel 8
Wallace, J. Bruce 6
Warrington 53, 180
Washington 53
Webb, Beatrice 7, 31
Webb, Sidney 6, 7, 30
Wellingborough 56, 67
Wells, H.G. 30
Welwyn Garden City 36, 45–7, 55, 59
West Coast Main Line 140, 168
White Paper 1965 172
White Paper 1997 115
Whiteway 79–80
Wild Mammals (Hunting and Dogs) Bill 200
Williams, Aneurin 32
Williams-Ellis, Clough 74–5
Winstanley, Gerrard 207
Wise, M.J. 85
Women in community ventures 80–1
Workplaces 23
Wormwood Scrubs 129
Wythall 56

Zehlendorf 90
Zetetical Society 6

229

Index compiled by Geoffrey C. Jones

Learning Resource
Centre